Naturekind

Naturekind

LANGUAGE, CULTURE
AND POWER BEYOND THE HUMAN

MELISSA LEACH

JAMES FAIRHEAD

PRINCETON UNIVERSITY PRESS
PRINCETON & OXFORD

Copyright © 2025 by Princeton University Press

Princeton University Press is committed to the protection of copyright and the intellectual property our authors entrust to us. Copyright promotes the progress and integrity of knowledge created by humans. By engaging with an authorized copy of this work, you are supporting creators and the global exchange of ideas. As this work is protected by copyright, any reproduction or distribution of it in any form for any purpose requires permission; permission requests should be sent to permissions@press.princeton.edu. Ingestion of any IP for any AI purposes is strictly prohibited.

Published by Princeton University Press
41 William Street, Princeton, New Jersey 08540
99 Banbury Road, Oxford OX2 6JX

press.princeton.edu

GPSR Authorized Representative: Easy Access System Europe - Mustamäe tee 50, 10621 Tallinn, Estonia, gpsr.requests@easproject.com

All Rights Reserved

ISBN 9780691270678
ISBN (pbk.) 9780691270661
ISBN (e-book) 9780691270722

British Library Cataloging-in-Publication Data is available

Editorial: Rebecca Brennan and Rebecca Binnie
Production Editorial: Ali Parrington
Jacket/Cover Design: Hunter Finch
Production: Erin Suydam
Publicity: William Pagdatoon and Kathryn Stevens
Copyeditor: Francis Eaves

Jacket/Cover image: Valedi / Shutterstock

This book has been composed in Arno

10 9 8 7 6 5 4 3 2 1

For all the more-than-human inhabitants of Flints Farm,
past, present and future.

CONTENTS

Preface and Acknowledgements   ix

1	Introduction	1
2	Language and Culture Beyond the Human: Rethinking Communication Across Naturekind	16
3	Chickens	54
4	Horses	64
5	Plants	78
6	Bees	90
7	Bats	102
8	Forests	114
9	Soils	131
10	Seas	149
11	Cities	163
12	Conclusion: A Political Naturekind	177

Notes   205
References   225
Index   249

PREFACE AND ACKNOWLEDGEMENTS

As anthropologists we attend to the ways the world is lived by those who don't share the assumptions about it that we were raised on. One of us grew up a Londoner and the other on a British farm, trained to maximise whatever could be produced on the land, or extracted from it. We presumed in our youthful insularity that what we learned at home, with our peers and as students at our great universities were the correct ways of envisaging the world. In our agronomy and geography degrees we learned a lot about 'natural resources', and then afterwards and with all the curiosity of our generation's lingering coloniality, we were both drawn to study anthropology, or rather 'social' anthropology. As an academic discipline, social anthropology was perhaps confused as to whether it was the most humanistic of the social sciences or the most scientific of the humanities, but was happily secure in being about the entire 'human condition'. In taking a holistic approach to the study of human social life, it left the rest of 'nature' out of this whole. Our condition and our social worlds extended to humans alone.

As we embarked on fieldwork, we were intent on achieving the kind of double consciousness that so many have had thrust upon them, and so Melissa went to live in a Mende-speaking village in Sierra Leone, and James to a Kinyarwanda-speaking part of the Democratic Republic of Congo, before we united to live in the forest region of the Republic of Guinea. What we found at every turn were worlds misunderstood. People said to be ignorant of crop diseases nevertheless had sophisticated ways to ensure crop health. Those reprimanded for impoverishing their soils had all the while been enriching them, and this over generations. Those castigated for deforesting their lands had instead been nurturing forests on them. And hunters vilified for reckless killing had been thinking their way mimetically into their prey and in vocalising with them could not deceive without respect.

As we carefully documented how those we laboured with worked with their soils and vegetation, and how herders, hunters and fishers alike understood the animals on which they depended, so we came to appreciate local ecological knowledge, experience and perspectives on these worlds, casting these as hypotheses with which to confront existing sciences and build new ones. Our trilogy of co-authored books—*Misreading the African Landscape*; *Reframing Deforestation*; *Science, Society and Power*—probed West African forests in this way, while a suite of further articles and volumes did so for soils and carbon, biodiversity, and ecologies of health and disease. These works revealed the analytical and corrective power of local knowledge and experience for understanding ecological processes. This was not science evaluating Indigenous knowledge, but Indigenous knowledge evaluating and correcting the orthodoxies in regional scientific literatures. In this, we could hardly not be political. The assumptions and framings held by so many scientists and policymakers in positions of authority and that were so disparaging of the wisdom of those we lived among could only be understood as emergent from relations of production of knowledge inherent to colonialism and post-colonial globalised knowledge orders. And the effects were often brutal, from the death penalty applied to local fire management practices in 1970s Guinea, to the 'green grabs' that have dispossessed people of rights, lives and livelihoods in the name of nature protection ever since. Ecology, we learned, is political, and respect for radically different ways of knowing can stimulate productive conversations between scientists and citizens that reshape farming, forestry and conservation for the better. They can also remain divided in contested narratives about nature shaped by and perpetuating deep power divides.

Were those in the communities who corrected our ignorance and whose ideas we then probed properly recognised? We hope so. Grappling with the histories and legacies of extractive modes of research has been explicit in all we do: in our writings and in the documentary film *Second Nature* that we created with community members in Guinea in which it is they who speak their truths to power. We have been relentless, even repetitive, in our broader analysis of the way power–knowledge relations actually shape the truths that prevail in all corners of the sciences we encountered, and this has been core to the research centres, programmes and courses we have engaged in with colleagues, partners and students from around the world from our home and long-standing institutional base in Sussex: James in the University's Anthropology Department, and Melissa in the Institute of Development Studies.

The trouble as we began this book, however, was that all this critique did not resolve an enduring double consciousness—an enduring cognitive dissonance—and it was not one restricted to our experience as anthropologists, but also closer to home. Our own social worlds—our relationships, affection and grief—extended to our companion cats, to the donkeys, horses, sheep and chickens we nurture in our smallholding and to the wildlife and birds that live there too. We found ourselves loving the trees we planted in a patch of field and the soils we encouraged for vegetables in ways that went well beyond their physical and chemical composition. Why ever did we think our social worlds stopped with the human? Why ever did we not question the 'anthro' of anthropology? Might a social faux-pas, or a gesture or a word out of place not be part of the fabric of wider nature? A misplaced word or gesture can bring us trouble not just from our human friends, but also from the equines or chickens we live with; and according to those we lived with in West Africa, such faux-pas can bring trouble to fields, crops and wildlife too. What is social and moral—what is communicative—does not stop with the human, but overflows the species line. Thankfully we found many academics and public intellectuals taking this path in 'post-humanities' and 'more-than-human' social sciences, whom we encounter in this book, and we launched ourselves and our students onto this wave and have been surfing it for a while. They, in turn, have been drawing on traditions of thought, whether the ideas of their informants or of Indigenous scholars, and living amongst communities around the world, that have not been configured by Enlightenment science and education. Yet the same old frustrations lingered. Do these dissident insights from our anthropological fieldwork, from everyday life in our smallholding and from our academic co-travellers also have corrective implications for the biological and ecological sciences? How far might insights from these disciplines about what it is that shapes human social, cultural and moral life, and from Indigenous and everyday more-than-human life itself, extend to reconfigure paradigms in the natural sciences? Should the critique of human exceptionalism that is emergent in the social sciences extend into what we began to call 'naturekind'? As ever, we turned to debates in vast scientific literatures and encountered a veritable revolution in the way communication and culture are being thought about across the biological world, but, crucially, one facing emergent and fundamental contradictions. Extending social science insights beyond the human, as we do in this book, can begin to resolve these.

This is thus not the book that we expected to write at all, but writing it has been a pleasure nonetheless, and it is a book that, as ever, we have conceived, grappled with and written as joint and equal co-authors. It is inspired by those we have lived among, by our colleagues at Sussex whom we have worked among, and by the expanding and wondrous coterie of researchers worldwide who have been probing the biological and human sides of naturekind. We thank Princeton University Press and our editor Rebecca Brennan for their creative support in bringing the book to publication, and the anonymous reviewers for their time, critique and insights. We thank our readers, too, who also might not be of one accord but who, we hope, might see the sense in laying out a book such as this. As partners in life, we thank all the co-habitants of Flints and Flints Farm, past, present and future, including our ancestors and our children Cassie, Rory, Xanthe and Cesca, for inspiration, joy, love and hope for the worlds they belong to and will shape.

Naturekind

1
Introduction

The premise of this book is that we humans are losing our capacity to communicate and socialise beyond the human. This is in part because we are annihilating our cohabitants on this planet, but this book is not about that, or at least not directly. What concerns us is that too often we think that we do not communicate beyond our species, and that we cannot—that we do not share common languages and cultures with the world we inhabit—and so our aim is to profile how in fact we do so, and why this matters for our capacity to live well both with the non-human natures with which our lives and futures are entwined, and with each other. Just who this 'we' is also matters, as none of this is common to us all. Yet shared attention to interspecies communication, interspecies languages and indeed to interspecies cultures is crucial in order to contemplate together the sorts of action and power that are needed to secure life on a shared planet, in what we shall be proposing as a political naturekind.

The world is a communicative place entwining us with animals, plants, fungi and bacteria; but how to understand this? Many peoples allow that waters, rocks and winds can enter these conversations too, but should this be marginalised as quaint belief, or rather made central as a source of inspiration, acknowledging the many ways of knowing and being across a 'pluriverse'; the many worlds within a world?[1] This chatter and its signifying sounds, scents, touches and glances has been attenuated in the humanities as in the sciences. Those who speak about and research ecosystems, for example, capture how 'organisms' and the non-living world interact as a system, and admit that biotic and abiotic components are linked together through nutrient cycles and energy flows, but all too frequently ignore communication or envisage it as similarly functional, without 'meaning'. If there is life in ecosystems, it is almost zombified life. A forest becomes a sort of complex automaton; a massive 'Heath Robinson' machine into which the sun shines and rain pours, and out of which timber, carbon, biodiversity and other 'ecosystem services' flow. A sea becomes a marine system in which we might lament the decline of fish stocks and the choking by plastic bags, but this is a pollution that interrupts nutrient flows and poisons water to limit productivity, not one that interrupts communication and kills meaningful life. A world of conversations is erased in just the same way as sentience in animals is denied by the idea of 'instinct', and in plants by the idea of 'tropism'. Since the Enlightenment, anyone who craves to foreground meaning in conversations that extend with life beyond the human, and to find meaning immersed in landscapes, has easily and endlessly been dismissed as romantic; as adhering to something unscientific—beyond science; and as forlorn and kooky, succumbing to anthropomorphic self-delusion in face of Enlightenment rationality. In this book, we turn that science and its logics onto itself to allow more meaningful communication back in.

The last few decades have witnessed major advances in biology, ecology and the technologies for their study, precipitating what some now cast as a biological revolution, transforming understandings of animal and plant communication. This now unveils not just communicative capabilities and sophisticated signalling, but more 'linguistic' dimensions to life beyond the human. It reveals, too, that all life is cultural, because social learning conveys different ways of being down generations of the same species: from whales to fruit flies; from sage brush to slime.[2] For many biologists, 'culture' now supplements genetics as a 'second inheritance system'.[3] Evidence of emotional and moral intelligence is now being found across the animal world.[4] Hitherto unimagined connectivities are becoming apparent, between trees, plants and fungi, and wider life.[5] All this is being achieved with fast-developing theories of cognition and of communication as signalling—but without a theory of meaning.

Meanwhile equal advances over the last century in understanding human culture, linguistics and communication have been made across the human social sciences, especially in *anthro*pology. Theorising meaning is central to these, with attention to the study of meaning in sign systems more generally ('semiotics') and in language ('semantics'). The scope and ambition of these advances have, however, been restricted to the human world, on the presumption that humans are qualitatively different from the rest of life in terms of modes of communication, language and culture. This human exceptionalism has supposed that what we theorise for 'us' in relation to meaning has little bearing on understanding non-human nature and peoples' communicative encounters with it. Yet the new biology now reveals that this supposition can no longer hold. The aim of this book is to unshackle the 'social' science of meaning from its assumptions of human exceptionalism and expand its scope, showing its significance for social and cultural worlds beyond the human and for interspecies conversations and culture. What can a theory of meaning born of the social sciences offer to the study of linguistic and cultural lives beyond the human at the frontiers of biology?

Others share this agenda. A discipline has emerged that is concerned with precisely this—with theorising the making of meaning (semiosis) beyond the human—identifying itself as 'biosemiotics'. The trouble, as we outline in the next chapter, is that this discipline's edifice of reasoning is entirely premised on human exceptionalism, presuming that only humans 'have language', and that only humans 'have culture'—just at the time when many biologists are showing linguistic qualities in non-human communication, and the centrality of social learning and culture across the living world. To understand the exchange of meaning beyond the human, the discipline of biosemiotics has thus constrained itself to focus on supposedly 'non-linguistic' (or 'prelinguistic') ways of exchanging meaning. The discipline that aims to provide a theory of meaning-making beyond the human thus has one theory for humans and another that embraces the rest; one for 'us' and another for 'them'. By showing why such human exceptionalism is wrong, we can help refocus biosemiotics on insights from human linguistics and semiotics from which until now it has set itself apart. By pursuing an argument that the study of human meaning-making need not be so restricted to humans, we hope to develop and provide useful theoretical insights both for biologists and for social scientists, and indeed for all those humans whose social worlds have become so restricted to themselves, so isolated, so separated from more entangled, more-than-human life.

Darwinian evolution provides one theory for all. Yet both the social and natural sciences continue to make assumptions of categorical separation in the field of communication, language and culture, retaining a pre-Darwinian distinction, usually supposing that such separation is self-evident, empirically supported and a product of that very evolution. Such confidence now rings hollow, not least since findings have emerged that fruit flies have different cultures, that bees teach each other skills, that cockroaches easily learn to distinguish one person from another and that trees warn others to prepare for pestilence. The making and exchange of meaning is central to all life. It is life. How does it work? This somewhat experimental book attempts to develop and deliver a new paradigm for communication across 'naturekind', the term we choose to capture the communities and cultures through which humans are inescapably interconnected with wider life.[6] If it does not succeed entirely, it should at least lay out firmer analytical terrain for others.

Communication Beyond the Human

Our inspiration in developing this book was initially the strangely titled *How Forests Think*. In this work, anthropologist Eduardo Kohn reminds us that language of the sort that linguists study is just a tiny dimension of communication, even though philosophers and theologians have attributed to it enough substance to separate humans from the rest. He makes the simple point that if we paid more attention to the non-linguistic modes of communication that we share with plants and animals, then we would make better headway in appreciating how all life is embroiled in a complex unfolding of meanings.[7] As Kohn observes, for too long the study of communication has focused on what differentiates humans from other beings, not on what unites us: that is, shared ways of otherwise conveying meaning. Kohn's insight is that we cannot focus only on the linguistic abilities of particular species, including our own, if we are to consider communication across species.

The agenda is powerful in intent but more problematic in substance. Kohn had turned for his inspiration to a tradition of semiotics that was developed by a nineteenth-century American philosopher, Charles Peirce, who allowed that although meaning can be conveyed through symbolic forms of communication (classically those in language in which the meaning of signs is arbitrary and depends on convention), meaning could also be conveyed in what he and followers identify as 'non-linguistic' or 'prelinguistic' signs. Whereas in symbolic communication all signs carry arbitrary meaning, and users thus depend on knowing how to encipher and decipher their code, non-linguistic signs are envisaged as conveying meaning more directly, without recourse to such code, whether though their iconic relations (in which something conveys meaning through its similarity, as a portrait links to a person) or their indexical relations (in which something conveys meaning directly, as smoke links to fire).

As this tradition appears to allow meaning to be communicated without recourse to preestablished codes, and to communicate in the more subliminal, experiential ways that music or cinematic imagery might be thought to evoke, it allowed analysts of communication in such media to stray from theories that treat all communication as in some way language. Peirce's hypothesis was equally attractive to those who studied how meanings are conveyed beyond the human, in particular Thomas Sebeok and his followers, who developed the discipline of biosemiotics on the same assumption: that meanings could be conveyed in prelinguistic ways, avoiding pre-established codes.[8]

Those studying biosemiotics were drawn to this agenda in light of their comprehensive readings of twentieth-century biology, from which they concluded that only humans 'have language' (in ways that we shall discuss later, in more detail) or 'have culture', and thus inferred that the study of biosemiotics beyond the human could not be rooted in the study of human linguistics and linguistic-inspired semiotics. Since then, however, the digital revolution and the methodological revolution in biology that it has enabled have challenged this reasoning, and so much so, we shall argue, that appeals to non-linguistic signs and the making of meaning entirely independent of symbolic orders can be called into question.

Whereas Kohn wanted us to focus on the non-linguistic signing that he considered would be shared with wider nature, we are going to suggest what amounts to the opposite view: that there is no end to language and that there is no such thing as entirely non-linguistic signing in the making of meaning. It is to things linguistic that we must look to see what we share with wider nature, after all, not to things prelinguistic. This, however, forces us to reconsider what language looks like, and how to study it.

Our way forward will be to clean up and repurpose a rusty old theoretical tradition rooted in the structural analysis of language developed by Ferdinand de Saussure more than a century ago that most of us had shelved in the archive of the history of ideas. From the 1950s it was developed by Roland Barthes and others in the structural semiotics they applied to film and media studies, and by Claude Lévi-Strauss and followers in the structural anthropology they used to study (human) society. In this book, we question whether their approaches should have been limited, as they were, to human communication, and what they might offer if we apply them to the making of meaning across naturekind. By overcoming what we suggest have been missed opportunities to integrate insights from structural social science traditions, we forge a 'structural biosemiotics' that can provide a unified paradigm for understanding meaning-making across naturekind. This draws back into focus ways of knowing, being and theorising that have hitherto been marginalised—whether in Indigenous societies or, indeed, in the disqualified everyday experiences of all.

Ecological Connection in More-than-Human Worlds

The context that makes such an exploration important is the imperative for ecological connection in current worlds. New generations are expressing a palpable desire to 'reconnect with nature'. The urge towards immersion in forests, seas and wildernesses has given rise to industries. Health services have adopted 'green prescriptions' for both physical and mental health. It is not just city dwellers and parents with children who pace park paths knowing the importance of 'connection', without quite knowing what this might be, and who return home exercised and exhausted, perhaps, but somehow unfulfilled. Is this desire something instinctual in us, as some have argued: an innate human urge to 'connect' with other living beings and the assemblages of nature, an urge now characterised as 'biophilia'? Others argue that connection is missing in industrialised, urbanised ways of life, and needs to be rebuilt—through 'rewilding' ourselves and our ways of being. Connection, thus viewed, is a vital counter to the forms of separation and violence that are devastating non-human life on the planet we all share.[9]

But what is this thing, 'connection', and how should we understand it? To connect in this way is to appreciate and take part in the chatter and conversations going on in non-human natures, through which they are alive, and lively. It is to make and experience meaning from seas, mountains and woodlands. It is to admit relationships beyond the human back into communicative socialising and communities. It is not that humankind is separate from nature, needing to reconnect, but that we are all, already and always, part of naturekind. The problem is less about releasing an innate, inner biophilia, or about reconnecting, than about understanding and appreciating the inevitable myriad ways humans are already connected, while learning from those, including Indigenous peoples, for whom such interconnection has never been in doubt. The premise of this book is that ecological chatter has always been fundamental to everyone's life, and the to and fro of our friendships and enmities, struggles and relaxation. The dissatisfaction that some feel as 'disconnection' is, then, a discomfort at the shrinking of more-than-human gossip; the shrinking of social worlds as people annihilate their companions or come to live, produce and consume in ways that seem more separate from them. A desire for connection generates a modern myth that palliates those who think the problem lies within themselves, to be solved by a stroll in the park or hike in the wilderness, and not in the structural and discursive forces that produce separation and exploit, destroy and divide humans from wider life on a shared planet.

To the twentieth century problem of the colour line we must thus add the twenty-first century problem of the nature line. To racism and social discrimination we must add speciesism and anthropocentrism. And these are interconnected, inasmuch as worldviews that separate humans from non-human nature, placing them in a hierarchy, are bound up with those that consider some humans inferior ('less human') than others.[10] Entire research, educational and religious structures have worked powerfully to draw a line, separate and divorce humans from nature, and the so-called 'humanities' and 'social' sciences from more-than-human weddedness. The foundational assumptions of twentieth-century sociology, anthropology, economics and politics presuppose that the rest of life lies outside the human social order, to be studied as the stuff of zoology, botany, ecology and ethology. The 'environmental humanities' and 'environmental social sciences' that cross over the divide are still niche, inhabiting that tiny overlap in a Venn diagram, swamped by all that is left out. To speak of naturekind is to undermine the foundations of powerful institutional edifices; but can we not speak of it, nevertheless? What gives it substance? What weapons can be wielded in its cause?

Many authors across the environmental humanities and social sciences discern the same problems as we do and prescribe related, but different, medication to integrate the study of human worlds with those of other species; but few grapple with communication, and of those that do, very few with biosemiotics. In many ways this broad issue is as old as academia itself, dating from Pythagoras's school, which conceived of the animal and human worlds as inseparable, was concerned that human beings might be reborn in other animal life forms (in the tradition of metempsychosis) and eaten, and so promoted vegetarian lifestyles.[11] In early modern times, the French philosopher Michel de Montaigne allowed that all species might have their own languages and societies and castigated humans for their vanity in ever thinking otherwise.[12] Early in the Enlightenment, Baruch Spinoza dissolved distinctions between people and things animate, discerning the vital forces inhering in all.[13] In the nineteenth century, anthropologist Lewis Henry Morgan railed against the concept of 'instinct', as it erased the calculative capacities of animals, reducing them to living meat.[14] More recently, feminist science studies scholar Donna Haraway extended the emotive nature of human kinship, whether in love or grief, to her dog companions and a full spectrum of 'critters', undermining the foundational distinction between human society and other life, and thereby severing the Gordian knot that bound up all 'social' sciences together as separate from the rest of the life sciences.[15] Yet for Pythagoras, the issue was reincarnation across species, not communication. Montaigne acknowledged the languages of different species, but not communication between them. Spinoza focused on the inherent powers at work in encounters and alliances beyond the human, not their communicative qualities. Morgan recaptured the calculative intelligence he thought God had instilled in all species, but ignored their communication. Haraway recaptured the affection and sociality across species, but did not probe the processes of communication on which they might be predicated.

More recently, many authors in the social sciences and humanities have been drawing on this same genealogy to conceive of human worlds as part of wider living and non-living worlds. They question the validity of fundamental boundaries between humans and non-humans, and probe how these have been constructed. Along the way they have coined new terms and phrases that attempt to dissolve the boundaries in both science and the society and life it studies, albeit from their own, rather different, interests and angles. Thus Bruno Latour and colleagues have encouraged thinking and practice around actor-networks of human and non-human 'actants' (although not 'communicants') in integrated 'nature-cultures'.[16] Donna Haraway conceptualised the human and non-human as inextricably linked, making kin through networked, 'tentacular' practices (tentacles but not tongues and multimodal equivalents).[17] Deleuze and Guattari theorise 'assemblages' and rhizome-like connectivity (but not communicativity) across human and non-human life.[18] Anthropologist Tim Ingold describes 'meshworks' of dwelling (but not the chatter between dwellers).[19] Political theorist Jane Bennett assembled rubbish and chemical effluents together with the people, animals and plants entangled with it, exploring them together as 'vibrant matter', doing away with foundational boundaries between the living and non-living (but underplaying the communicative nature of such vibrancy).[20] In the 'cosmopolitics' of philosopher Isabelle Stengers, a multitude of beings—human and non-human, living and non-living—form a collective society (but little related to things linguistic).[21] And in the philosophical works of Karen Barad these entanglements extend inside human and non-human bodies too, in what she terms 'intra-action' (but not intra-semiosis).[22] Each of these writers offers theoretical insights that extend the social beyond the human; but the contribution of each can be inflected further by focused attention to communication.

More generally, anthropologists have pioneered 'multispecies' ethnographies that attend to human–non-human relations and the varied ways these are conceptualised and experienced across the world,[23] in works sometimes characterised as 'post-human' or as concerning social relations 'beyond the human'. Examples focus on social relations between people and various kinds of life and things, from horses and birds to mushrooms and mountains,[24] in settings that range from oceans and mangroves to high-tech urban and industrial landscapes.[25] Geographers and cultural theorists similarly grapple with such 'more-than-human' or 'post-human' relations—with wildlife, with disease, with urban environments, and much more.[26] Historians document past understandings of the natural world as being so vital and alive that animals such as pigs and weevils could be prosecuted in court. Several Indigenous scholars and philosophers have been pioneering these traditions, raising voices and developing critical theories from worlds that never posited the kind of human–non-human boundaries and hierarchies that have for so long dominated scientific canons.[27] They identify and condemn the subjugation—albeit only partial—of these ways of knowing as a feature of imperialism and colonialism.[28] The works of such authors mesh with the direct self-expression and claims of self-identified Indigenous peoples and local communities, enunciated in social movements that manifest themselves, often exuberantly, at global encounters. They mesh, too, with a flourishing worldwide popular environmental literature, fiction and arts practice indicative of a veritable zeitgeist, thriving in media and exhibitions. Our concern is to probe more systematically the communicative dimensions, specifically, of the more-than-human entanglements addressed in these traditions.

A few works have done just this, crossing social and biological science boundaries to focus explicitly on trans-species communication, and, with the same intent as ourselves, to theorise communication beyond the human. The present volume is a contribution to an emerging dialogue, aimed at securing further its analytical foundations. Anthropologist and natural historian Gregory Bateson was prescient in probing multisensory messaging across species, but drew distinctions between linguistic and prelinguistic communication that we will call into question.[29] Vinciane Despret's 'philosophical ethology' also attends in detail to the everyday multisensory ways by which humans and other animals communicate in co-becoming, but considers human linguistic and semiotic concepts as overly anthropocentric—as human-derived concepts that undermine her project of allowing non-human animals to reveal their agency and be interesting on their own terms.[30] Philosopher Eva Meijer, writing on 'when animals speak' and looking 'toward an interspecies democracy',[31] describes all interspecies communication as 'language', yet without directly addressing semiotic theory. Only a few authors have engaged explicitly with debates on communication beyond the human that are unfolding in the discipline of biosemiotics. One is anthropologist Eduardo Kohn, as previously indicated, and another is philosopher Dominique Lestel, whose agenda is to 'think together' human and animal societies, urging study of the complexities of trans-species communication and 'hybrid communities'.[32] While Kohn focused on so-called prelinguistic communication as uniting all life, Lestel, in his pioneering synthesis of biological and anthropological approaches dating from 1998, argues, as we do, that 'culture' must be conceived of as a semiotic phenomenon in animals as in humans; since then he has been probing eco-semiotics and 'hybrid human/animal communities of shared meaning, interests, and affects'.[33] Our structural-biosemiotic paradigm shares, advances and provides further analytical grounding for this broad agenda, urging, in the light of new developments in biology, that the principles developed in structural semiotics be extended from human life to all life. 'Language' as usually conceived—as a human thing centred in speech and writing—thus comes to be construed as a small subset of much larger communicative and cognitive orders extending across multispecies worlds; that is, across naturekind.

Researching Communication Beyond the Human

In this book, then, we probe wider communication through debates that have been had about how human languages convey meaning and about how human social orders are communicative orders. We take these out into naturekind, investigating communication amongst a wide range of animals, plants, microorganisms and the ecological assemblages they comprise and inhabit. Our focus is less on how non-human beings communicate amongst themselves (although we do consider this), and more on how humans join this chatter; on conversations between humans and non-humans, both in the more dyadic relations that people establish with non-human companions, and in the wider assemblages in which all live together. Expanding theories of semiosis from human language and culture to consider these beyond the human might be construed as 'anthropocentric', but so might the critique of it, presuming things linguistic to be only human.

It takes time to develop communication with other beings, as a Saami reindeer herder theorised for the TV camera:

> I was out herding with one of the elders. We came to a river we had to cross which was running high so he said to his lead reindeer, 'Let's go.' He said it a few times and then they went across. He told me we should talk to the reindeer. 'They understand,' he said. But now in this motorised age, people are in such a rush. They don't have time to talk to the reindeer.[34]

Cognitive and communicative framings develop as part of the flow of life and relationships, of doing things together. It takes time to listen to the reindeer, but not just time, as how communication develops depends on the river crossings and how they go. A good crossing, and both sides learn something about the gestures, the pulls, the sounds that helped. A falling-in or a near miss might bring different signals—those of fear or relief, perhaps. But something is learned, and both reindeer and herders change, if subtly. Communication thus develops—emerges—through activities, practices, actions, events, as do the connections so forged. Who has the time to codevelop such communication across species? Many who live with companion animals are aware of this problem, and regret they are too busy to develop communication potential with them, distracted by the rush of work and activity. Yet people do develop such languages when lives and livelihoods depend on it—when they depend on those reindeer, on cavalry horses, on sheepdogs who communicate with sheep better than human farmers do; on the elephants who built the ancient cities of Cambodia.

If this is the case, then it might not be philosophers and linguists who will guide the way into understanding communication beyond the human, especially not those with disciplinary archaeologies rooted in a radical conceptual separation of humans from non-humans. Nor can our methods rely on the repertoire honed by biologists and ecologists—hypothesis-driven experiments and observations, refined by technologies, applied in the physically or conceptually controlled settings of labs and fieldsites. Equally we cannot rely on the conventional methodological repertoire of (human) structural anthropology and semiotics, with its interviews and observations by an external researcher.

Instead we must turn to those whose everyday worlds extend beyond the human realm; to the experiences and insights of those actually living and communicating with non-human natures. In many ways, they must be our teachers. Methods are needed to enable their perspectives and experiences to shape the research, and to apprehend the complex webs of meanings and practices that emerge through more-than-human life. Such methods fall within what has come to be called 'multispecies ethnography', a methodological repertoire that takes the principles and practices of the anthropological stock-in-trade of ethnography, with its emphasis on careful, detailed listening and (participant) observation, into worlds beyond the human. It becomes important to participate in encounters across species, listen to accounts of them and discuss soundscapes and scentscapes, and more, using an array of multisensory approaches.[35] The multispecies ethnographies we carry out ourselves and draw from others in this book make use of, and contribute to, this expanding methodological repertoire.

Since we argue that communicative encounters beyond the human are part of everyone's everyday lives, everywhere, our foci and sites could be infinite. Those we have selected illustrate a range of kinds of human–non-human communicative interaction: with particular animals—chickens, horses, bees and bats; with trees and plants; and with assemblages of living and non-living entities—forests, seas, soils and cities. For each, we draw on ethnographic studies from a range of different locations. Some are auto-ethnographic, drawing on our own encounters beyond the human in the United Kingdom; some derive from our earlier work as anthropologists in West Africa, which we now revisit and re-view through a multispecies communicative lens; and some is from the work of other ethnographers, accessed through their published and online works. The scope thus extends from the South Downs of Sussex in the UK to the mountains of Peru and Guatemala; from North American woodlands to Pacific seascapes, from homesteads in the rural Philippines to rooftops in urban Pakistan, New York, and beyond.

Communication and Power in and for Naturekind

The way we treat the world around us in farming, fishing, industry, construction or everyday life alters how we think about it. Equally, the way we think about 'nature' alters how we treat it—our relations of care and respect, of extraction and pollution, or of disregard or neglect. Questioning human exceptionalism on the basis of the evidence generated by the biological revolution is thus not simply an academic question, as there are much wider interests and issues at stake with which these findings and interpretations are entangled. Whilst engagement with questions of language beyond the human is prompted in part by the new biology, it necessarily raises political, economic and juridical questions, as it alters whether we conceive wider life to be a resource, like iron ore or timber, or our kin, as part of our own social and cultural order.

Thus whilst this book addresses contradictions that are now emergent from laboratories and scientific fieldsites and develops a paradigm that can resolve them, it has far wider implications for existing economic and political orders and the way these treat the more-than-human world. Since these are the same consumptive orders that now threaten life on our shared planet, a science that reframes language and culture also opens up—and gives impetus to—the prospect of a new environmental politics: what we must call a 'political naturekind'.

One way of appreciating how ideas about 'nature' relate to wider economic and political practice and experience is to recognise a series of interlocking separations that underpin and help reproduce current human exceptionalism, whether experiential and spatial, economic, or conceptual. Firstly, we can speak of the extinction of experience of more-than-human worlds, whether linked to their annihilation, spatial segregation or conceptual separation (human exceptionalism). Many human social worlds have already experienced environmental apocalypse—in the collapse of biodiversity, in the impacts of climate change, in the devastation of landscapes and waterscapes. For many more, this is to come. Economic forces precipitate destruction, with non-humans exterminated, incarcerated, exploited or commoditised to serve the modern human industries feeding unrestrained consumptive desires. The extinction of experience is linked not only to this absolute decline, but also to a decoupling of modern lives from the variety involved in direct experience beyond the human. Modern livelihoods are increasingly divorced from, not wedded to, non-human natures, given the urbanisation, industrialisation and agrarian mechanisation that attenuate any enduring form of everyday encounter with them. The economic forces that alienate people from the land also produce radical inequalities of multispecies encounter, whereby for many the costs in time and money reduce opportunities for the more-than-human sociability on which the lives of the marginalised once so heavily depended. Not everyone can afford to keep pet animals or travel to wildlife-rich places;[36] though to be sure there exist too counter-tendencies to the above generalisations, whether it be marginalised migrants turning to picking matsutake mushrooms in Oregon, or impoverished urbanites finding solace, in spite of all, in blasted late-industrial cityscapes.[37]

It is not just that in industrialised worlds people no longer live lives in such close connection with non-human natures. They may be actively kept away, as conservation policies divorce people from lands and seas set aside in the name of 'nature', weighing heavily on those displaced whether by national policy or the economic accumulation of conservation territory by elites. Ideas and practices that associate 'nature' with a pristine, unpeopled wilderness, or that attempt to 'wild' and 'rewild' places by letting nature 'take its course', can similarly instantiate separations of humans from nature.

As politico-economic forces drive loss of experience of mutual interdependence this in turn becomes an integral driver of extinction itself, because without such experience, decline may be neither noticed nor much cared about. Marine scientists have coined the term 'shifting baseline syndrome' to capture the realisation of specialists that new-generation scientists grounded their understanding of 'normality' (of the quality and variety of marine life) on their experiences early in their own careers, failing to perceive the degree to which their baseline in this respect is already impoverished as compared with the generation before. As this process has continued for generation upon generation, it is shocking now to read documented evidence of the extent of more-than-human life in antiquity.[38] As biologist Carl Safina puts it, '[a]nimals, plants, habitats and human cultures vanish [...]. Even the memories of them are disappearing.'[39] The decoupling of lives from non-human natures amplifies this shifting baseline of experience.

Economic separations are furthered by conceptions of living beings as 'natural resources', reducing life to things that exist for instrumental human use and justifying domination over them. Conceiving of more-than-human life through the lens of its 'species' and 'species diversity' can play into such economic separations: it is too easy to view an individual creature, an elephant or a jaguar perhaps, simply through the lens of its 'species', as if one individual is just like another—an exemplar of the species, each as substitutable for another as knives and forks in a table set; as for any mass-produced commodity. This deferment of individual life to the species is ever more prevalent as species become commodities in approaches that trust that in trading nature we can save it: in emergent policies and applications such as biodiversity credit markets, natural capital accounting and nature-based financial and business tools, within the wider field of capitalist conservation. Biodiversity 'offsetting', for instance, licences the killing of newts in one place, so long as other newts are invested in somewhere else. Actual life, in such thinking, does not matter: only 'species' or 'units of biodiversity'. It is as if someone could substitute a lost friend by meeting another of 'their culture'; or as if someone could kill their companion cat before a holiday only to buy a new one on return, if it were cost efficient. The reduction of trees and forest patches to 'carbon' in carbon accounting—a commodity to be exchanged in carbon and offset markets—follows a similar economic logic, with a similar effect. In such a conceptualisation something is ruptured, and it is that something which we probe in this book; it is something communicative, and it is real. It is why we don't keep trading in our pets—it is at the heart of affection. Whilst species characteristics surely shape communicative encounters and possibilities, and while those interested in these encounters talk of 'interspecies' interactions and communication, as we will also do sometimes in this book, we also need to acknowledge relations and communications that are more personal and inter-individual.

To the ecological, economic and political forces that cascade together to support human exceptionalism can be added religious ones that foster worldviews founded on human exceptionality and the ability to conceive of social and cultural worlds as exclusively human. Again, such worldviews unleash people from any limits to the kinds of human expansion and the restraint that comes from respect.[40] Conceptual separations are upheld, too, by the institionalisation of international scientific disciplines that have historical genealogies rooted in theological presupposition: the disciplines shaping research which divide so-called 'natural' science disciplines of biology, ecology, botany and so on from the sociology, anthropology, political science and economics that study human-only society and culture. The upholding of problematic conceptual separations extends to those authoritative philosophies of mind and consciousness that draw hard distinctions between these and the 'body' or the 'subconscious', finding in them the locale of reason, self or soul. Such claims carry weight even though the language, communication and culture they draw on even to express their concepts inevitably transcend 'mind' so conceived, and even though so much of our experience and what we communicate with each other, and more widely, is embodied, unintentional, subliminal and 'subconscious'. The analysis of communication should not be confused with theories of mind. When it is, the particular exceptionalities of the human mind are extended to become the basis for a more general exceptionalism, overlooking how all beings have their own exceptionalities.

These forces of separation—economic, spatial, conceptual—thus interlock, sometimes in a veritable cascade that has been permissive of the destruction of worlds in which people and non-human natures thrive together; even of ecocide. The new findings emerging from the biological sciences are therefore revolutionary not only in a methodological sense, emerging from the technological revolution, but also in the politico-economic sense of revolution, by necessitating the paradigmatic reframing that we conduct here, which disrupts this cascade. Our reframing aligns with a rather different cascade of concepts, experience, ethics, and economy, focused not on human–nature separation, but on inseparable entanglements, connections, mutual interdependence and care. The concept of 'naturekind' carries, appropriately, a double meaning, connoting also living with kindness (itself etymologically linked to 'kin-ness') towards the more-than-human world of which all are a part, aligning with concepts and ethics of care,[41] and those seeking to understand and advocate for them. Living well together will be impossible without a greater degree of affection, conviviality and care, to drive political engagement; affection that is itself a product of sociality beyond the human.

Our particular contribution here is a focus on communication, and a structural analysis of it. To build this agenda, the next chapter develops a new paradigm for communication across naturekind based on structural biosemiotics, which we then ground and elaborate in subsequent chapters. Chapter 2 thus brings into dialogue the biological evidence for linguistic communication and cultural practice across naturekind, with insights about these phenomena from structural anthropology and social semiotics, taking these beyond their human-exceptionalist origins in a discussion that for a few pages is necessarily conceptually dense. The chapter thus lays out the fundamental theoretical contribution of the book. The following chapters bring the arguments outlined to life, probing communication beyond the human in everyday settings across the world. From chapters 3 to 7 the focus is on companionship and the dyadic communication that humans codevelop with particular beings—whether in cohabiting and spending time together, work, sport, or a host of other activities and interactions. We move from chickens to horses, plants, bees and bats. Then from chapters 8 to 11 the focus is on wider assemblages of humans and non-humans, living and non-living things. Here there are multiple ongoing conversations between animals, plants and their surroundings, and as humans enter or hack into them, they enter lively, communicative worlds, with many significances for all parties. Our chapters here consider how this happens in forests, seas, soils and cities. Each chapter tells particular stories about how communication interacts with connection and emotion in entangled lives, how this challenges the separations that dominate in science, policy and politics, and the implications for pressing questions concerning environment and health.

In the final chapter, we show how a focus on communication brings into sharper resolution what is at stake in some of the most intense discussions of our era about planetary predicaments and futures. We sum up what previous chapters have shown about communication beyond the human and its implications. We show how theory and practice in environmental politics, too, has been dominated by a focus on the only-human, limiting strategies for a more recoverable earth, or transformations towards greater mutual flourishing. We offer a new theorisation of a political naturekind, involving communication beyond the human as deliberative environmental politics, and considering how non-human beings and assemblages might 'speak for themselves' in political processes, not just be spoken for. And we show how appreciating communication beyond the human can contribute to more hopeful narratives, caring relations and forms of recovery in which all life might thrive together; a politics of and for naturekind.

2

Language and Culture Beyond the Human

RETHINKING COMMUNICATION ACROSS NATUREKIND

This chapter develops a new line of approach for understanding communication across naturekind. The prevailing idea that humankind can be distinguished from wider life on the basis of qualitative differences in language, culture and the making of meaning is crumbling, and an adequate theory of communication across naturekind requires a new synthesis across the biological and social sciences.

First we illustrate the veritable revolution in the biological sciences—supported by new technologies—that now finds that many kinds of animal communicate 'linguistically', including insects and, in some respects, also plants. Such communication occurs across a whole spectrum of senses: from sonic modes including ultrasound and infrasound, visual modes including ultraviolet, infrared and bioluminescence and tactile modes including pressure, to heat, electromagnetism and an infinite variety of chemical mixes. There are presumably other sensory modes yet to be discovered. This biological revolution shows that core aspects of language once thought uniquely human, including symbolic signs and syntax, are at play in wider naturekind. This revolution also reveals a more pervasive transmission of behaviour via social learning, including of communicative behaviour, such that some biologists now call social learning or 'culture' a 'second inheritance system'. Sperm whale cultures vary by clan, songbirds learn dialects, and fruit flies transmit mating preferences in culturally different ways across generations.

The biologists making these discoveries have managed until now with very limited attention paid to the questions of 'meaning' that are so central to all debates concerning human language and culture. Most probe communication through a narrow model of cue and response, and of the outcomes of communicative signalling in terms of functions and evolutionary fitness, in explanations that are thus ultimately mechanical, or 'mechanomorphic'. While they may infer that meanings are being exchanged as part of social learning, they have no theoretical apparatus to elaborate how. It just happens. This position is increasingly precarious. It is not just that if social learning and culture are a 'second inheritance' system, then questions of meaning exchange become crucial to evolutionary theory. It is also that we might wonder where non-humans' learning and culture ends, and ours begins, and how exactly to understand interspecies communication, and participate in interspecies culture. In short, we need to understand more fully how meanings are made and shared across naturekind.

A discipline has been developing that aims to study exactly this, calling itself biosemiotics, and since the 1970s it has been extending the study of semiotics—the long-standing study of meaning-making in people—to the wider biological world. As we consider in the second part of this chapter, however, the approaches taken have been rooted in the very assumptions of human exceptionalism that biology now challenges. They have been probing communication in wider naturekind only through so-called 'prelinguistic' signs that somehow allow for communication 'directly' without symbolism, without the establishment of learned codes or conventions, which until now have been presumed to be distinct to humans. It is this that can no longer be upheld given the new findings from the biological revolution. Once we have outlined the issues involved here, we turn in the third part of the chapter to key insights from the social sciences and humanities concerning language and culture that have hitherto been restricted to people, and show why and how these can be extended into the study of the making of meaning and communication in naturekind—and thus also of people's communication beyond the human. It is this agenda and the questions it raises that form the focus of our subsequent chapters, which—as we indicate more fully in the final part of this chapter—explore a diversity of communicative communities across a pluriverse of life and experience.

Scotland, 1492

In the summer of 1492, King James IV of Scotland ordered that two newborn babies should be taken to the windswept and rocky island of Inchkeith across the choppy waters of the Firth of Forth near Edinburgh. They were to be raised in isolation by a deaf-mute woman until they 'came of proper age' and learned to speak. The king or perhaps his courtiers wanted to know what language the children would speak; what would be the natural, God-given language of humankind that would emerge from their mouths, unpolluted by the prevailing languages of the realm. The results remain unpublished, but as the children reputedly began to speak Hebrew we might cast some doubt on the veracity of the whole account, especially as the island was repurposed as a syphilis colony within four years and this particular experiment was likely curtailed.[1] Other monarchs through the ages have been drawn to conduct the same type of experiment to identify a God-given human language, but with contradictory results. The Egyptian pharaoh Psamtik, for example, found that children deprived of language began to speak Phrygian. Similar investigations initiated by the Holy Roman emperor Frederick II, and by the Indian Moghul emperor Akbar, by contrast, both turned out to be inconclusive.

Nowadays we find the idea that there might be a single, innate human natural language to be absurd, but strangely we do not find this same presumption absurd for other species. We presume that animals, and perhaps even plants, communicate in some innate way that is specific to their species; that dolphins somehow 'speak dolphin', whales whale, nightingales nightingale, and toads toad, and that armed with a recorder and Artificial Intelligence we might one day perhaps 'decode' what they are saying in their 'natural' language. Even if we rip individual organisms out of their experiential eco-social and spatial worlds, place them in a cage or breed them in the lab, we presume we could still access 'their language'. But is this not somewhat presumptuous?

A Biological Revolution

Recent research in the biological sciences, building apace on longer-established studies, has finally begun to render this presumption absurd. As those investigating animal cognition and communication deploy newly possible sensory technologies, they reveal example after example that force us to question why we have come to think about human language so very differently from the way in which we consider modes of communication in the wider living world.[2] Whilst this research has focused principally on communication within particular species, we probe it here to open up approaches to communication beyond such limits, be it across species or between beings and their surroundings, and to establish the linguistic principles involved.

A spider wandering across its web might once have been assumed to be doing just that, moving randomly or perhaps instinctively in seeking food. Yet biotremologists using lasers to detect vibrations now reveal that spiders live in a web of signals; they tune each thread differently, pluck and bounce on them, and quiver (tremulate) their abdomens, communicating almost musically across the webs. They produce a field of vibrations in what researchers call 'stereotyped sequential displays' or 'structured signalling' that carry meaning the extent of which eludes us, but in ways in which the significance of their varied ordering of signs (syntax) is evident.[3]

Honeybees had long been understood as exceptional after Karl von Frisch showed in 1967 how they convey information to each other about the direction and distance of nectar sources, and indeed many other things, through their waggle dancing, in which patterned movements symbolise distance and direction and the order of signals (syntax) alters the meaning.[4] Their signalling now turns out to be more than just a dance, as the head-butts and the airflows created by wing beats also transmit information: for example, in correcting misinformation about nectar sources when there is none left in the flowers.[5] What was exceptional about honeybees was that we humans came to understand their 'language' earlier than in the case of other species.

We now find that the living world communicates across a vast spectrum of possible senses, and that it does so with symbolism and with syntax. Academic journals and more popular works now bulge with evidence of modes of communication far beyond human perception, or even beyond our imagination, amongst a vast array of beings, that just a decade ago was almost inconceivable.[6] Biologists are going beyond the documentation of simple cues—indicators of information which elicit a response—to explore signals and signalling systems embedded in evolution, a signal commonly being defined as 'any act or structure which alters the behaviour of other organisms, which evolved because of that effect, and which is effective because the receiver's response has also evolved'.[7] Signals are carried across a vast range of visual and sound modes, from the ultrasound of bats to the infrared of bees. Communication through sound extends through vocal signals, the underwater whistles of dolphins and the songs of whales, to the sounds of the beat of wings. In crested pigeons, for example, specially modified wing feathers produce distinct sounds that act as alarm signals during escape flights, and listeners flee on hearing them.[8] Spiders can pick up sounds in the air using their webs as acoustic antennae, and because spider silk responds so precisely to vibrating air molecules, these are among the most sensitive 'ears' in the natural world.[9] Calls can convey information through tone of voice as much as content, allowing penguin chicks, bats or lambs to find their particular mothers, for instance, by discerning the subtleties of one voice from the general cacophony. In large penguins, this involves combining the two distinct sound frequencies, or 'voices' that they can produce in a beat pattern which varies between individuals, and which mates or chicks can identify.[10]

Communication through touch has long been understood as central to relationships amongst social animals. Thus primates use touch, often combined with gestures, in the elaborate and flexible communicative strategies that are important in caring for infants, creating and maintaining groups and hierarchies, and organising activities such as travel and foraging. Grooming—removing lice, parasites and other objects from the fur with hands or mouth—does not just clean, but forges social bonds; chimpanzees who regularly groom other individuals are more likely to be included in food sharing.[11] Yet tactile communication has now been found to exist amongst a far wider range of animals, and beyond adjacency. Fish, for example, sense and interpret water pressures around them through the lateral lines along their bodies, with such 'touch at a distance' providing enough acuity to sense neighbours and coordinate movement in shoaling.[12]

Electrical signalling is important too. Mormyrid fish communicate at night by constantly emitting and perceiving brief electrical pulses, with the particular waveform and intervals transmitting information about the sender's species or group identity and current behaviour and motivations. Pairs of fish synchronise their electrical pulses to each other, and echo their responses to address specific individuals and establish brief pairwise communication—conversation—within a group.[13]

Chemical communication between animals has long been known about, with many releasing pheromones in urine or from scent glands to mark territory or signal to others. Ring-tailed lemurs even have 'stink fights' when, during conflicts, males rub their tails across their wrist and chest glands before waving them at each other.[14] Now it appears that trees too communicate with each other through chemical signals, released into the air, soil, or through their symbiotic underground fungal networks. The emerging evidence, while still controversial, extends to their communication about droughts and pest attacks.[15] When a tree is under attack by insects, it can release volatile organic compounds into the air, alerting neighbouring trees to the threat.[16]

Such communication integrates multiple senses (or 'modes'). The displays of flies when courting combine visual, vibratory, acoustic and chemical signals, and frogs when mating combine calls with vocal sac display and water surface vibrations; in both cases the message conveyed depends on the combination.[17] The latter example integrates the visual, sound and vibration physically, in ways familiar from human speech: as we humans speak, we both move our lips and utter sounds, and looking at the moving of lips alters the sound (the sign) that we 'hear'. Other multimodal combinations, not so integrated, include the courting behaviour of flies, and the combination of displays and vibratory songs of jumping spiders, the communicative effect of which lies in coordinated sequences.[18] This is little different in principle from the gestures and movements that politicians combine with the rhetoric of their words, or the PowerPoint displays that support lecturers in getting their meaning across.

Biological sciences increasingly reveal the symbolic character in communication beyond the human, and beyond the bee.[19] Biologists and psychologists are increasingly encouraging researchers to look for evidence of the two key characteristics of symbolic communication: arbitrariness, whereby a signal is neither innate nor directly linked to its referent, but has to be learned, and conventionalisation, whereby learning involves the regularities and norms of useage common to a group.[20] Many animals, from dogs and chimpanzees to dolphins, have been trained to use arbitrary symbols.[21] Capacity for symbolic communication has been shown experimentally, as in the classic example of vervet monkeys who produce three distinct alarm calls in response to different predators: when these calls were replayed in the absence of the predator, the monkeys responded in a predator-specific way, suggesting that the sound functioned as a symbol of the predator.[22] Symbolic communication is harder to identify in the wild, yet evidence for it is mounting for cetaceans, apes and more. Thus the songs of sperm whales convey individual and group identity,[23] whilst signature whistle exchanges are a significant part of a greeting sequence that allows dolphins to identify members of their group when encountering them in the wild.[24] In the Tai forest of Côte d'Ivoire, chimpanzees drum on trees to convey information, symbolically, on travel direction.[25]

Evidence is also mounting of the significance of sign order: thus of syntax, a key feature of language. Those drumming chimpanzees in Côte d'Ivoire alter the particular pattern of beats, and with it, the message conveyed. Many primates combine vocal signals such that the message of the whole depends on the particular integration of the parts, according to rules of combination; of syntax, that is, or 'grammar'. Chimpanzees in the Budongo Forest of Uganda have a 'pant hoot' which signals who they are, and when they feed in the company of a powerful friend they often make food calls, but they can also combine these calls (almost as words) to attract others in their social world—who get angry if they are not told—to the food. The combined message, eliciting a particular response, is built from the sum of the parts.[26] Putty-nosed monkeys have two acoustically distinct alarm calls that when combined convey something else entirely. A *hacks* signals an aerial threat such as an eagle, and a *pyows* signals a problem on the ground, but in combination they convey 'Let's go,' even when unrelated to predators.[27]

Digital analysis of bird calls and songs similarly shows how sequences of signals matter in relation to the responses of those listening.[28] Pied babblers appear to put words into sentences as they alert others to low threats with one sound, and call others to join their travel groups with a different sound, but combine these calls to gather and mob a threatening predator on the ground.[29] Birds of the tit family combine sounds in ways that elicit different responses in listeners according to how they are arranged: Japanese tits mob a predator when they hear either their own alert and then a recruitment call, or those of other species in mixed flocks, but do not respond when the sequence is reversed. Such compositionality in language works across bird species, and in using it they 'extract a compound meaning from novel call sequences using an ordering rule.'[30] Similarly, when chestnut crowned babblers utter two sounds, the sequence matters. One arrangement stimulates nestlings to beg, but the other way round it coordinates movement. Research grapples with whether such 'phrases' acquire sense from their constituent parts or are more 'idiomatic'—as in a human phrase such as 'kick the bucket', in which the sense of the whole is not tied to any of its components.[31]

Where sounds are seemingly repetitive, different messages can be conveyed according to the precise repetition, speed and rhythms. Thus for colobus monkeys, roaring in long bouts at long time intervals warns of eagles, whereas short bouts and intervals (introduced by a snort) warns of a leopard. Mexican freetailed bats click singly when investigating novel stimuli, but in bouts when socialising with others, suggesting this repetition conveys a different message.[32] Studies of birdsong syntax consider how notes (or syllables) are assembled to produce songs and how songs are assembled into sequences. Song sparrows, for example, sing a variety of song types which they repeat in a consecutive series or 'bout', going cyclically through their repertoire. They vary the order from cycle to cycle, with the choice of the next song shaped by what they have sung in the past nine or so cycles. Yet whilst song sequences clearly follow syntactical rules, it is unclear what the variations convey.[33]

Examples illustrating hitherto overlooked linguistic capacities in nature continue to multiply. They suggest that an absence of such powerful evidence across a wider spectrum of life may be an indicator less of any lack of such capacities, than of our inability to discern them—as yet. Such 'linguistic' communicative abilities are exceptionally difficult to demonstrate. The several works that purport to show the absence of syntax and language in life beyond the human turn out to have been significantly flawed. Recently the methods that have been applied to analyse vocalisations in wider life, and which find no evidence of discrete segments and syntax, have also been used to investigate humans, and find, absurdly, that people too have neither syntax nor language.[34] The methods are not powerful enough. This underlines the methodological challenges involved in determining the existence of grammars even in streams of sound, let alone in other modes and across them.

Methods, however, are advancing apace, and as they do, they reveal complex signalling and syntax in ever more species and in ever more complex forms. Technological advances in Artificial Intelligence, in particular, have begun to discern increasingly sophisticated patterns and their variations in streams of sound.[35] Algorithms investigating the songs of birds now show how sounds that are not adjacent nevertheless relate to each other, suggesting a much more complicated syntax. Thus they can find that 'syllables' sung by a canary influence the choice of another syllable, five syllables later.[36] They find discrete segments in streams of parrot song, and patterns in the complexity of their enormous combinatorial variation that are akin to those found in human speech.[37] Yet what these human-like vocal components actually convey remains unknown; their communicative content remains largely unfathomable. Unable to fathom meaning in such streams of sound, some researchers write off the content as meaningless—as 'bare sound', finding significance only in the fact of singing (not its content), and how it conveys something about evolutionary fitness.

Syntax in human language is found not only in combinations of signs, but also in how combinations are themselves ordered in relation to other combinations in a hierarchy and through recursive relationships, and this provides human language with enormous power. Syntactic recursion enables humans to create infinitely long and novel utterances through multiple embeddings, as in 'I rode the horse that belongs to my sister who trained him in the style that she learned from her friend'. Such recursion has long been upheld as a feature unique to human language, but despite the almost impossible methodological challenges, there are indications of both hierarchy and recursion beyond the human. For example, humpback whales sing songs with nested multi-level hierarchies, in which individual sounds are arranged in 'phrases' that repeat to form 'themes' that are sequenced in song cycles of four to seven themes, often lasting for ten minutes or more. Male singers usually share a song pattern with their particular population, and as this changes incrementally over months and years, and occasionally transforms completely, all singers adopt the new pattern.[38] Capacity for recursion has now been shown in rhesus macaques,[39] while studies of chimpanzees now show how they combine and recombine vocal utterances to produce very diverse and flexible vocal sequences that could support numerous differentiated meanings. Orangutans produce long calls in which sound sequences are recursively nested within each other.[40] European starlings seem capable of distinguishing patterns that rely on recursion,[41] and recent studies show that crows perform it on a par with human children, producing long recursive sequences involving deep embeddings.[42] Nevertheless, given that this evidence of recursion cannot be tied to any meaning, it can always be contested on the basis that the researchers are just proving animals' abilities to learn inconsequential patterns.[43]

Just as human communication draws on signs from the world around, so biological studies also show the importance of context in non-human communication: how the information conveyed by signals can derive not just from their combinations, but also from how they relate to the surroundings. When some Campbell's monkeys in the rainforests of Côte d'Ivoire shout their *krak* warning of predators, aerial and terrestrial, those listening infer whether the problem is from sky or land depending on where they are in the landscape.[44] Chimpanzees make similar 'pragmatic inferences' when the significance of their signalling takes into account their place in the social hierarchy, or other past experience.[45] It is signalling in relation to context that allows several species to deceive others with false alarms.[46] For example, an insectivorous African bird, the fork-tailed drongo, makes alarm calls that it also deploys in non-threat contexts to scare others away and steal their food.[47] Such deception depends for its effect on evoking a context that is actually not present.

This biological revolution extends to a massive growth in the documentation of 'animal culture'. Many animal behaviours, including communicative ones, once thought to be innate to 'a species' are now understood to be acquired through 'social learning'. Communicative and other practices thus differ between groups within a species, shaped by the particularities of their upbringing and experience. Culture is 'a major part of what whales are', with behaviour varying by clan and accounting for their historically situated dietary traditions, hunting practices and song genres.[48] The feeding and movement patterns needed to live well in a particular place are not just responses to the specificities of the place itself, but are cultural and inherited. When sperm whales sing, their song sequences or 'codas' convey not just who they are, but which group or 'clan' they belong to—a clan being an intangible concept. As Carl Safina has put it, '[a]s boundaries of human social groups can be marked by language differences, sperm whale society boundaries are reflected by differences in group codas'.[49]

Many biologists now refer to social learning and 'culture' as a 'second inheritance system',[50] and this concept is not restricted to 'the usual suspects'—the 'more intelligent' primates and porpoises—but extends 'down the phyla', to invertebrates, taking in birds on the way. Thus Amazon parrots in Costa Rica use three regional dialects in their calls: north, south and Nicaraguan. All are of the same species and gene pool, but they learn the dialect culturally, amongst those they are raised with.[51] Likewise parakeets of the same species show different call dialects between the different European cities they are colonising.[52] Different groups of the same species of sparrow learn different dialects in their songs.[53] Amongst many songbirds, songs are learned by imitating specific tutors.[54]

Egyptian fruit bats learn their vocalisations too, partly through the particular sounds of the crowd they are raised with, but also directly from their mothers.[55] Sac-winged bats of the Americas speak to their pups in 'motherese'—a characteristic 'educational' language that is part of their upbringing and acculturation.[56] Many fishes 'acquire dietary, food site and mating preferences, predator recognition and avoidance behaviour, and learn pathways, through copying other fishes'.[57] Insects are good learners too—honey bees' waggle dance may be innate, but the exact dance patterns and what they convey are learned as young bees copy older ones.[58] Fruit flies transmit mating preferences culturally across generations.[59]

Social learning may also extend to plants. Plants interact through a great variety of chemical molecules in liquid or gaseous forms. Plants have usually been understood to respond to 'tropisms' that presuppose perception, signal transduction and response, whether to gravity, light, the sun, the moon, oxygen, chemicals, electric fields, magnetic fields, temperature, touch or wounding, and probably more besides. Yet the concept of 'tropism' in the plant realm is as mechanistic as the concept of 'innate' or 'instinctual' in the animal kingdom: invoking the kinds of clockwork mechanical automata manufactured during the Enlightenment era as a metaphor for understanding wider life, in what others have called 'mechanomorphic' reasoning. Increasingly, however, researchers have found that plant perception and communication is not just a mechanical response and exchange of chemicals (of 'information'), but involves 'the active coordination and organisation of a great variety of different behavioural patterns—all of which must be mediated by signs', and which are combined through syntax and in particular contexts.[60] Studies of sagebrush plants show how they acquire repeatable individual personalities, differing in response to the 'alarm calls' that some emit to others when attacked by herbivores.[61] Plants that have already experienced particular pathogens or herbivores recognise the signs of such enemies more rapidly and react more effectively. Plants can thus remember past events, shaping their subsequent responses; thus they 'learn', and 'have memory'. They also exchange information with each other, and with microbes, herbivores and herbivore enemies.[62] Some have been found to distinguish self from non-self, so that if parts of self (a root, for instance) are separated, they become non-self, and the plant will respond differently to those parts (in reacting to a root exudate, for instance). Some plants pass memories down generations through maternal or epigenetic mechanisms with signs being picked up by progeny differently according to inherited parental experience. Such examples suggest that 'cumulative culture', accumulated as transmitted from one generation to the next, once thought unique to humans and now found across many non-human animals, may feature much more commonly still in nature.[63]

The Need for a Theory of Meaning in Biology

The biologists making these discoveries have usually got on fine without probing how the signals and their syntax which they are revealing actually carry information; how they convey meaning. Most biologists speak of 'cues' or 'signals' without worrying how, exactly, a signal 'means' something to the cockroach, pigeon or whale they experiment with or study. As biological studies of communication have progressed and its linguistic characteristics have become clearer, however, this lacuna has become increasingly apparent to a few biologists interested in its wider evolutionary significance.[64] Here we consider how this contradiction between methods and findings manifests in the biological sciences, and the limits to the ways most biologists approach meaning, before considering approaches in the parallel field of 'biosemiotics', and then in the human social sciences, to help resolve it.

In mainstream biology, 'meaning', to the extent that it is addressed at all, is usually deduced through the responses of study subjects. Many experiments rely on teaching subjects to respond to particular 'cues' and so rely on cross-species social learning, but even this rarely prompts questions concerning how such learning happens and what its implications are. It has been possible to conceive of the animals simply being programmed (or 'conditioned') to respond to cues, as if the learning mechanism involved was uninteresting. The biological concept of signalling is defined in terms of evolutionary biology, with signals—and even complex combinations of signals—being identified through their effects in altering the behaviour of other organisms, but with both signalling and effect seen as an outcome of evolution—as functional to evolutionary fitness. Communication through signalling is thus rendered entirely mechanistic. Even recently, as the significance of syntax has become so clear, there has seemingly been little need perceived to dwell on what, exactly, the content of messages is, and how information is actually exchanged; on how signalling and syntax convey meaning. They just do. Biosemiotics scholars Kalevi Kull and colleagues are amongst those puzzled by this silence, arguing that biologists who adopt such an approach must have 'an asemiotic conception of life as mere molecular chemistry', and yet they use terms that invoke meaning, such as 'information', 'learning', 'fidelity', which they apply 'in an allegedly metaphoric way'. This permits an implicit assumption that such behaviour 'can be reduced to mere chemical accounts if necessary'.[65]

Biologists have been led to this neglect of meaning by 'the most awesome weapon in animal psychology',[66] which has come to be known as 'Lloyd Morgan's canon'. This dictum, coined in the colonial, Victorian 1890s, states that "[i]n no case may we interpret an action as the outcome of the exercise of a higher psychical faculty, if it can be interpreted as the outcome of the exercise of one which stands lower in the psychological scale."[67] It is the Occam's razor of animal inquiry. Simpler, 'parsimonious' explanations are always those that are more 'mechanical' or more 'imitative', somehow programmed 'through repetition', and are less 'cognitive'. Students of animal cognition are taught to avoid considering whether 'behaviour is driven by internal representation'[68]—sometimes referred to as 'modelling', 'thinking', or 'conceptualising'—and to limit conclusions to simple, parsimonious explanations which refer to 'behavioural rules' and the kind of regularities that statistics can discern. The study of animal cognition and communication (let alone that of plants and other forms of life) is thus always to be understood in the most mechanical of ways possible.

We can appreciate the deficiencies of this canon by turning the dictum upon our human selves. It would be parsimonious and adequate for biologists who adhere to it to interpret observations of prisoners regularly walking clockwise around an exercise yard at the same time every day in terms of mechanical social learning. But what of those prisoners' interpretations and reflections on their incarceration? What of the power relations and personal histories that got them there? A whole genre of bioanthropology once flourished by looking at people in such mechanistic ways, neglecting the sense they make of their worlds and actions. Now accused of dehumanisation (and indeed coloniality) in the social sciences, the canon is alive and well in biology.

Even where behaviour beyond the human is, necessarily, in some way interpretive ('hermeneutical')—when for example, a beaver must calculate which way to fell a tree to create a dam, and how to achieve it—Morgan's canon produces a kind of 'mechanical hermeneutics'; a conditioned response to the signals around. Ironically enough, it was another Morgan, the nineteenth-century anthropologist Lewis Henry Morgan, who produced one of the earliest studies of animal meaning-making. His *American Beaver and His Works* of 1868, shows these animal architects involved in sophisticated interpretation and communication.[69]

The very research that is now revealing signalling and syntax across naturekind, and the significance of social learning and culture to it as a 'second inheritance system', has thus until now either avoided the thorny question of 'meaning', or approached it in very limited ways. Whilst researchers might recognise that the communicative element of a cue or signal depends on its ordering (syntax), context (pragmatics) and learning (history, memory), these aspects are not considered as tied together in an exchange of meaning (as they would be with people), but simply in terms of behavioural outcomes. Students of animal communication and cognition who follow Lloyd Morgan's canon face accusations of 'mechanomorphism', but most remain unconcerned, as their vantage point seems to have worked well enough until now, and retort that anything else would risk 'anthropomorphism'. It would succumb to 'a new wave of romantic biology'.[70] Those defending themselves in this way, however, now face a novel and insurmountable problem. The magnificent biological revolution conducted according to these principles has produced a set of internal contradictions and challenges, which cannot be pursued without a more elaborated theory of meaning.

A first challenge concerns evolution. If social learning and culture really constitute a 'second inheritance system', then they will need to be fully included within evolutionary theory. Processes of evolution cannot be understood without attention to their modes, efficiencies, limits and so on, which, once considered only genetic, are now considered to include social learning and culture. These factors now account for some of the most interesting questions concerning evolution, yet the biologists who approach them are equipped with only a very limited conceptualisation of culture, focusing on socially learned behaviours and the outcomes of 'cumulative conditioning' of the kind which they can observe. These certainly fall short of social scientists' understandings, which place 'webs of meaning' at the heart of what they might call 'culture'. Upholding a divide between the webs of meaning shaping 'human culture' and the 'mutual, cumulative conditioning' shaping 'more-than-human culture' is based on assumption, not evidence, and can now be seen to be as flawed as the study of human society once was when some biological anthropologists observed the regularities of people's lives without paying attention to the categories, classifications and meanings that shape social practice.[71] Biologists will need to attend to and theorise meaning more deeply, as the social sciences have long done, if their inquiries into evolution are to advance.

Second, it is not species that live, but organisms, and they do not live alone. If a theory of meaning is important for understanding the culture of a given group of animals of a particular species, it is all the more important for understanding the intercultural worlds and assemblages in which organisms of different species coexist. The biological and ecological sciences to date have either studied 'interspecies communication' with the same mechanistic tools and methods used to analyse communication amongst members of a given species, or extended such relations through ideas of an 'ecosystem' in which organisms and non-living components are seen to interact in more or less systemic ways.[72] Until now, the conceptualisation of ecosystems in most contemporary ecological sciences has been as mechanical as the conceptualisation of their components, with, as yet, hardly a place for exchanges of meaning within such assemblages. But the very success of the biological revolution in revealing communicative, cultural life within species compels attention to these exchanges, between species and beyond species; without it, we will not be able to understand ecological dynamics.

Third, people are part of these assemblages and their dynamics.[73] The challenge is thus not just to understand communication amongst members of a given species, or just to understand communication amongst many organisms in assemblages they coinhabit, but also to grapple with how people, animals and plants are entangled in what Dominique Lestel characterises as 'hybrid' or 'mixed' communities.[74] As Lestel suggests, these are 'enchanted space of trans-specific communication', in which we need to comprehend our shared life and exchanges of meaning with those we dwell with.[75]

In short, if the biological and ecological revolutions are to continue, both at the level of communication amongst particular species and organisms, and at that of the wider assemblages of which they are a part, these disciplines will now need to embrace an understanding of how meanings are made and shared. Biologists often endeavour to avoid looking at living beings as though they were humans (characterising this as anthropomorphism), and sociologists and anthropologists have endeavoured to avoid looking at humans as though they were the kinds of beings envisioned in biology (characterising this as mechanomorphism). In this book, by approaching the study of meaning in a way that eschews fundamental distinctions between kinds of beings, we develop an analytical vantage point that overcomes this impasse; one that, in bridging biological and social sciences, will, we hope, be useful to both.

We should not confuse a theory of meaning with a theory of 'mind', or 'consciousness', or 'interiority', or 'thinking' or equivalent, not least because most meanings are communicated 'unconsciously'. Much of the meaning even that conscious human beings make and communicate is far beyond our consciousness, as a foundational experiment in animal cognition once neatly revealed. A horse aptly named 'Clever Hans' was said to be good at arithmetic. Researchers investigated, posing him arithmetical questions. The stallion signalled correct answers by tapping his foot the right number of times. Could a horse really solve arithmetical questions? Further research showed that this astute horse was all the while reading the researchers' own unconscious expressions and bodily gestures: and reading them accurately enough to decode precisely when he had tapped his foot sufficiently to give 'the answer' they wanted.[76] Hans was thus able to pick up on the unconscious expressions of several independent human researchers. He was no good at maths, to be sure, but very good at reading the meaningful but unconscious expression of human emotion. A theory of exchange of meaning is thus a very different thing from a theory of mind, cognition or consciousness, although it will clearly have something significant to contribute to the way we think about these. And we need not become embroiled in the debates that rage over questions of which non-human beings might be said to have consciousness, awareness of self and so on, since—as we will show—we can probe processes of meaning-making and sharing that are not predicated on assumptions about these things.[77]

Semiotics and Biosemiotics

If the biological and ecological revolutions are to continue, then, there will be a need to understand more fully how meanings are shared. Since the 1970s an entire discipline has been developing that aims to study exactly this, although, remarkably, it has remained very marginal to the practice of biology itself. This discipline of 'biosemiotics' has sought to extend semiotics—the long-standing study of meaning-making in people—to the wider biological world as 'the science of signs in living systems'.[78] The difference between 'signs', as used in semiotics, and 'signals', the apparent equivalent in the biological sciences, is more than terminological, involving distinct conceptual roots and assumptions. In this second part of the chapter, we consider the field of biosemiotics, but identify a fundamental problem with the approaches it takes: biosemioticians still consider humans as exceptional, as the only beings communicating 'linguistically'. They have been probing and theorising communication in wider naturekind at many levels from the cellular to the ecosystemic, but only through what they classify as 'prelinguistic' signs. In this way, they allow for communication of meaning in nature without recourse to symbolism, syntax and the establishment of learned codes or conventions, all of which they have understood to be distinct to humans. Yet it is this that can no longer be upheld, given the new findings from the biological revolution. The fundamental presuppositions of existing biosemiotics are thus flawed, and so here we develop an alternative. To do this we turn to traditions in semiotics that biosemioticians have rejected. The fields of semiotics and biosemiotics are conceptually challenging, and can come over as arcane, even impenetrable. Here, however, we must engage with their core ideas in a discussion that is inevitably dense at times, but necessary to identify foundational problems and ways to address them.

Semiotics became central to the humanities and social sciences, especially from the mid-twentieth century, but it has its roots in antiquity, as signs and runes have been thought about for as long as they have been read. The central problem, as we see it, was visible from the outset, when the fifth-century Berber scholar (and Catholic saint) Augustine of Hippo divided signs into two classes: 'natural' signs and 'given' (or 'conventional') signs (*signa naturalia* and *signa data*). 'Natural signs', he discerned, 'are those which, apart from any intention or desire of using them as signs, do yet lead to the knowledge of something else.' He uses the example of smoke that suggests fire. Conventional signs, such as words or flags 'on the other hand, are those which living beings mutually exchange in order to show, as well as they can, the feelings of their minds, or their perceptions, or their thoughts'.[79] Augustine's distinction is thus predicated on whether or not signs are intentional and conscious. He thus treats the countenance of an angry or sorrowful man as a 'natural' sign as it 'indicates the feeling in his mind, independently of his will'. According to these distinctions, when Clever Hans the horse was reading the comportment of those researching him, he was thus reading human 'natural signs', not those based on convention or shared codes. Augustine asked whether animals such as chickens respond to signs, and whether they do so intentionally or instinctively, but having raised the question, he sidestepped answering it, given that his interest was only in the human intentional signs important for Christian doctrine. Meanwhile his natural signs, on closer inspection, are not what we would see as natural at all, as they are predicated on historical experience. As Augustine himself wrote, it is 'through *attention to experience we come to know* that fire is beneath, even when nothing but smoke can be seen'.[80] It is through experience, too, that people and horses come to read countenance. So can there be such a thing as a 'natural' sign?

In the twentieth century, two traditions of (bio)semiotics emerged, one of which answered 'yes' to this question, and the other, 'no'. The one that allowed there could be natural signs, drawing on Augustine's distinction, has dominated modern biosemiotics and, indeed, debates in the social sciences that attend to what adherents consider as prelinguistic or phenomenological cognition and communication. We will need to outline the key features in this tradition so that we can better appreciate the problems that the biological revolution now poses to its fundamental presuppositions. We then turn to the other tradition, which claims there is no such thing as a 'natural' sign; a tradition that is rooted in the structural linguistics that developed from the foundational work on linguistics by Ferdinand de Saussure. We need to do this to show why Saussure's theoretical basis for a science of human language and meaning can be (and should be) expanded beyond the human: across naturekind.

Mainstream Biosemiotics and Its Problems

Those developing the tradition of biosemiotics have been hugely influenced by the work of Thomas Sebeok.[81] He argued from the 1960s that a theory of meaning in the biological world must be based on forms of signing that do not presuppose linguistic, symbolic conventions. Such conventions would require the establishment of codes shared across a semiotic community, which he thought was not to be found across nature.[82]

Accordingly, biosemiotics was founded on the assumption of human exceptionalism: that only humans 'have symbolism', 'have language' and 'have culture'. Sebeok posited that language 'appears with syntax', and after reviewing the vast literatures on animal behaviour he could then find 'no [...] syntactic structures' in animal sign systems[83]—a conclusion which, as we have shown, the biological revolution has since proved wrong. On the basis of twentieth-century biology, Sebeok thus argued that human language and culture are categorically discontinuous from the rest of life when it comes to the making of meaning.[84] Language had its own origin, and marked humans out as wholly different from the rest of naturekind.[85]

Semiosis (the exchange of meaning) beyond the human, Sebeok argued, would have to be based on 'prelinguistic' semiosis. He thus invoked the idea of 'natural' signs, dating back to Augustine, but which had been reformulated in the nineteenth century by the American philosopher and polymath Charles Peirce.[86] Sebeok drew from Pierce that it would be possible to distinguish such 'natural' or 'prelinguistic' signs from 'symbolic' signs that have convention-based relationships with their objects.[87] Biosemioticians thus came to focus on 'prelinguistic' signing, in which signs are considered to carry meaning more directly. Such prelinguistic signs might be 'indexical', whereby the signifier is linked to the signified through cause: for example, as a lion's roar might index a lion, or a chemical scent might index the flower that created it.[88] Other prelinguistic signs might be 'iconic', linking signifier with signified through resemblance: for example, as one male lion resembles another, or as the topography of mouths, noses and eyes of an angry face resembles that of another angry face, even of another species. This conceptualisation of prelinguistic signing thus imagines that there might be communication in nature without symbolism; without the establishment of codes or conventions which might be learned from each other or from experience. This, it was supposed, was distinctive to humans.[89]

Biosemioticians today continue to distinguish signs that are 'symbolic', which (most assume) are restricted to human language and culture, from those that are 'indexical' and 'iconic', which they consider as 'prelinguistic', and ubiquitous to life. They thus reproduce deep-rooted distinctions and separations between humans and the rest, between culture and nature, between the social sciences (humanities) and the 'natural sciences'; whilst all the while acknowledging that 'the human animal' might nevertheless have communicative continuity with the wider living world, should we focus on the 'prelinguistic' signing that (it is allowed) we have in common. If we want to foreground continuity beyond the human, they argue, we should focus not on symbolic language (that distinguishes us), but on the semiotics of natural signs (that unites us).[90] Some anthropologists, such as Eduardo Kohn in his book *How Forests Think*, have recently adopted this analytic when seeking to expand social analysis beyond the human. They downplay attention to things 'symbolic' and focus on things indexical and iconic.[91]

Yet as we have shown, as the biological revolution unfolds, it is finding the symbolism, syntax, language and culture in wider naturekind that the founders of biosemiotics neither expected nor made room for. Crucially, within Sebeok's paradigm, it is not that naturekind's linguistics does not exist, but that it cannot exist.[92] Before he died in 2001, Sebeok dismissed research into animal linguistics as 'unscientific nonsense', but the subsequent biological revolution has since revealed this attitude as mistaken. So whilst the study of semiosis in naturekind is of fundamental importance, the tradition of biosemiotics that claims to lead it is founded on an error. A way forward is to abandon the analytical presuppositions and associated lexicon of this tradition, and turn to a different one in the study of signs.

Building a Structural Biosemiotics

As the biological revolution reveals the significance of syntax and of culture across naturekind, we need to probe the potential of a second tradition in semiotics. This 'structuralist' tradition emerged from the study of how human language conveys meaning, initially in the structural linguistics of Ferdinand de Saussure,[93] which others later extended (in structural semiotics) to understand how images, film, music and all human multimodal media also convey meaning, and further (in structuralist anthropology) to show how regularities in human social practices cannot be divorced from the meanings they convey.[94] Here we consider whether these insights can be advanced beyond anthropocentric understandings of human communicative and social orders to naturekind. What opportunities arise? What theoretical challenges are raised, and can these be overcome?

Structural analysis proposes that signs always acquire and convey meaning in their relation to other signs, and not from any inherent, 'natural' connection between the sign's signifier and what it signifies. That relation is thus necessarily arbitrary and so is based on convention, not mechanism. The meanings that a sign conveys can derive from two kinds of relations with other signs: first, its sequential relation with other signs that are present, as in syntax (or grammar)—'it is green', for example, means something quite different from 'is it green'; and second, from its relation to similar but contrasting signs that are absent, but in relation to which it acquires sense. It is in relation to each other that words such as 'hit', 'beat', 'slap' or 'chastise', for instance, come to acquire their precise, differentiated meanings when they occur in speech or writing. Signs that express related concepts thus delimit each other reciprocally, in a mutually defining field of signs: a 'semantic network'. If the concept of 'chastise', in our example, did not exist, then 'all its content would go to its competitors'.[95] Conversely, as new signs (words as concepts) are coined—'domestic violence', for example—they must barge into an existing semantic network, gaining meaning in relation to the other concepts, but in ways that also affect, and refine, the others' meaning.

These structural insights were originally developed around narrow human language, but were developed beyond narrow language in 'structural semiotics'. From the 1950s Roland Barthes extended this to the study of meanings conveyed by images, film, music and dance, probing the multimodal making of meaning amongst humans, in 'any system of signs whatever their substance and limits; images, gestures, musical sounds, objects'.[96] Just as word order alters meaning in speech and writing, so the ordering of image sequences in film and notes in music, convey meaning (even though, as in language, we might not be consciously aware of this). Thus a film sequence in which an actor places a suitcase on a bed and opens it suggests that they are about to travel, unless they have just travelled, in which case it conveys that they are settling in. The meaning drawn from this syntactical relation between signs is also shaped by the semantic field that gives sense to the suitcase. The suitcase gains its meaning in relation to other types of bag that it is not (its semantic field) including the handbag, the rucksack, the swagbag and so on, which if substituted would alter the sense. We might perceive that the 'signifier' of the bag relates almost naturally to what it signifies, until we reflect that suitcases, handbags and rucksacks have not always been around and their significance is not universally understood in all corners of the world.

Structuralist anthropologists such as Claude Lévi-Strauss took these insights still further, into wider social life.[97] They explored how there are syntactical orderings shaping social performance, too: for example in rituals and all that might be considered ritualistic, as well as in everyday life. Thus there are grammars in how we behave at a meal: in which foods are served together, the sequence of expected courses, the passing of dishes, in who is served first, and so on. The meanings conveyed in a meal and how one behaves in that context also depend on its place in a wider field of meals (a breakfast, as opposed to lunch or dinner; an everyday dinner as opposed to a special one for a birthday or festival). The meanings of particular foods, or of what we dress up in to attend a meal, depend too on their place in a field of possible foods and possible dress. Wedding cake, birthday cake, fried cakes eaten by the roadside—all acquire meanings as part of the semantic field of cakes, and in distinction from each other. And whether one wears a dinner jacket, suit, boubou, jeans, shorts, party dress or sari carries meaning, each making sense in contradistinction to the others; meanings that become apparent when one 'gets it wrong', or indeed dresses out of line intentionally. Thus social life also evolves 'grammars' that everyone in a common social world performs in relation to, even though they can hardly articulate those grammars, and might not consider them to be 'language'.

Such simple examples can be multiplied infinitely, but whether in narrow language, multimedia or wider human social life, structural analysis reveals how signs convey meaning within structural orderings.

Structural approaches have faced many criticisms when used for the study of human social worlds. These will be pertinent, too, as we probe their potential for understanding how meanings are perceived and conveyed beyond the human, and we therefore raise questions with regard to these key foundational principles, below. We find, furthermore, that several aspects of structural approaches will have to be reworked if we are to rid them of problems associated with this tradition's origins in the study of humankind, and especially in the analysis of narrow human language. Addressing these issues will enable us to develop a structural biosemiotics appropriate to wider naturekind, that subsequent chapters illustrate.

A fundamental implication of the structuralist tradition is that any sign is not simply something separate that we use to express ourselves, as an instrument of communication, but that it aquires its qualities in relation to others: to the extent, indeed, that the very ideas that we might wish to express cannot exist apart from our language; or, if they exist, do so only in 'a form that may be described as amorphous'.[98] Structuralist analysis thus deploys not merely a theory of language, media and society, but also, and simultaneously, one of perception and of cognition. In its parlance, all signs that convey meaning unite a 'signifier' (such as a word, a film element of a bag on a bed, a slice of cake) and a 'signified' (its meaning). Perception occurs because the sign and its meanings are indistinguishable, as two sides of the same coin. Many critics have raced to reject this, arguing that most of what we perceive is 'prelinguistic': we perceive the taste of what we eat at that meal, the heat of the soup or stew, and likewise we perceive depth in the landscapes around us; we perceive love. Surely, critics argue, it would be ridiculous to suggest that all cognition and perception is entrapped by linguistic signs. Yet when we reject this on the basis that we experience direct, 'prelinguistic' perception, we miss a crucial subtlety that will become far more important as we expand the scope of structural analysis to wider naturekind; for so-called 'prelinguistic' perception itself follows structural linguistic principles. The problem is where we draw the line around 'language'. What is 'linguistic' has hitherto been too narrowly defined.

This problem can be probed in relation to claims that structural approaches are unable to deal with images that are iconic and indexical, in which the signifier relates 'naturally' to the signified. Gilles Deleuze is among those who argued that images were part of 'prelinguistic' communication, writing of how '[t]he cinema seems to us to be a composition of images and of signs, that is, a pre-verbal intelligible content (pure semiotics), whilst semiology of a linguistic inspiration abolishes the image'.[99] Yet this overlooks how all that we 'see' as images, and the music we 'hear' is itself sensed in relation to structural orderings. For example, the sensation of perspective in what we see derives not just from our binocular vision but, as Leonardo da Vinci theorised effectively so long ago, also from integrating a range of subliminally perceived contrasts in size, in light and dark, in colour and in distinctness, to which Rembrandt added variation in texture. We rarely notice how we integrate these fields of meaning,[100] but that we do so becomes immediately obvious when they are inconsistently combined. Thus, as da Vinci cautioned fellow artists, if you paint distant objects with too much definition, it confuses the mind and interferes with subliminal perfection in the rendering of perspective. The 'aesthetics' of images in our visual perception is thus grammatical enough to pick up when particular signs in their fields of meaning are wrongly ordered.

The structural tradition not only considers that the signifier and the signified are united, but also that the link is arbitrary. That is, there is no inherent relationship between them, only one established by convention within the code of a particular language. This tradition thus rejects the very idea of 'natural' signs: that what signs signify might acquire sense from some intrinsic connection to the signifier (whether iconic or indexical). Critics question whether signs can possibly be so arbitrary even for people, let alone for wider naturekind. Are some gestures not so hardwired into us that we communicate subliminally with such 'natural signs'—through a smile or a tilt of the head that conveys approval, for example? What about a dog wagging its tail in happiness, or a horse laying back its ears in aggression? What, indeed, of the above example of perspective? In all these, the meanings of such signs appear to be conveyed in an innate, not an arbitrary way. Yet we should not jump too fast to such a conclusion. Even an apparently 'innate' gestural meaning belongs to a field of meaning: a grimace, after all, is not a smile, and a horse's ears laid flat back against its head convey a different meaning from ears pricked forward, within the field of 'ear positions'. Thus even what is 'beyond our control' when communicating—what is outside of consciousness—still conveys its meaning as part of a structural order; it is one sign, not another, within a field of signs, and conveys meaning as part of an ordered grammar. When we do become aware of such gestures, we can play on them, even riff on them: as we put on an ironic false smile, for example, or as an actor mimics the stoop of despondency, or as a horse might even deceive its rider. If there are innate dimensions to signing, then these provide what we can more profitably consider as biological 'anchors': 'innate' anchor points that nevertheless acquire sense in mutually constitutive fields of signs. Such biological anchors, in being more innate in origin, might have a 'pre-coded' (prelinguistic) quality, but they only become meaningful signs within a structural order.

We can therefore allow that there are potentially three dimensions to signs: that whilst there might be innate dimensions, and whilst some signs may have indexical and iconic qualities, they nevertheless acquire meaning in symbolic orders. These are not qualitatively different kinds of signs. 'Consider a photograph,' as one explication succinctly puts it: 'it has properties in common with its object, and is therefore an icon; it is directly and physically influenced by its object, and is therefore an index; and lastly it requires a learned process of "reading" to understand it, and is therefore a symbol'.[101] All signs, however 'prelinguistic' they might seem, acquire meaning within symbolic orders and depend for their meaning on conventions within fields of signs and syntax. This analytic actually conforms to the original conceptualisation of signs by Peirce, who considered all signs to have three dimensions; but it differs absolutely from subsequent biosemiotics, which claimed to draw on Peirce's work but envisaged three qualitatively different 'kinds' of sign.[102]

The distinction between so-called prelinguistic communication and symbolic communication has therefore been overdrawn. Moreover, just as there is something symbolic in so-called prelinguistic communication, so, conversely, there is something prelinguistic in all symbolic communication, as the signs and syntax we use in language and life become every bit as unconscious as so-called natural signs. We are rarely conscious of the syntax we follow, or of the fields of signs that give sense to each. When we speak, neither are at play at a conscious level; yet both are essential for effective communication. To put it another way, our spoken or written language is embodied such that there is a direct, 'prelinguistic', quality to the symbolic signs we use. This is why words are so real—why they can hurt as much as a stick. Words can be physically shocking; poets and playwrights can evoke tears through the words they use. Many peoples around the world accordingly interpret words as being part of the fabric of the world. Symbolic meaning is thus no less embodied than are the more direct, indexical or iconic dimensions of signs. The same is true in the 'linguistic' structuring of our social lives, whereby the social signs and grammars we follow in our everyday lives are encoded outside of consciousness.[103]

Linguistic Communities

In the study of human language, the link between a signifier and signified has a conventional, learned dimension that has to be understood in relation to the history of the particular 'linguistic community' in which it arises.[104] Critics can therefore immediately pose the question: how can we establish a theory of meaning across naturekind that presupposes the existence of linguistic communities? To address this, we must consider how we understand ther term 'linguistic community'. Until now, theorisation has been based on exclusively human contexts, and analysis grapples with our socialisation into the arbitrary dimensions of our particular language(s). Yet we can also conceive of what such a 'community' might be in more-than-human ways: a community that unites animals with the people they live with, for instance, or with other animals of a different species. Even if many non-human organisms do not form obvious 'linguistic communities', it is still the case that all life occurs within 'communities' of a sort; within groups of animals or plants, in interspecies assemblages and within ecological niches. Within these, patterns of orderliness can be perceived, however emergent, contingent and shaped by the beings' particular context, needs and sensory system. So as long as there is a sensory system and memory that can make out an orderliness, then it is to this we should look to find a linguistic community configured by its common signs and syntax. Human language becomes a special case of this, one that recognises an orderliness based on speech and writing.

Indeed, as biologists now uncover the significance of syntax and culture across naturekind, we certainly have no reason to continue to suppose that communication according to such structural principles, and the development of codes and conventions, is restricted to humans. Structural principles can be extended to an understanding of wider, multimodal language, communities and cultures across species. Clever Hans the horse, as we saw earlier, was perfectly able to 'read' the subconscious signs of his human handlers and researchers who probed his mathematical acumen. Yet such skills are not to be found in all horses, so we can assume they have codeveloped between a particular horse and particular humans, producing what we can now understand as a kind of linguistic community—albeit a community of the few. And what is true for Clever Hans we can probe for across naturekind. As we shall be exploring in later chapters, those who keep chickens in their yards or train in equestrian dressage know that they do indeed codevelop signs that make sense in relation to other signs with their particular chickens or horses (not chickens and horses as whole species). They develop grammars that neither can articulate, but which shape their co-activity and the co-living and con-sociality it is embedded in. We follow Lestel in this, who thus argued for the study of interspecies cultures, urging that attention be paid to 'hybrid communities sharing meanings, interests and affects'. He, too, acknowledges that these 'hybrid communities are primarily semiotic communities', and that this must involve codes: 'it is precisely because humans and animals are able to share common codes that they can constitute shared communities'.[105] The questions become more interesting, not less, when we extend them, as we shall in the chapters that follow, to communication even beyond animals: with and amongst plants, other organisms and, indeed, to assemblages that combine all of these.

The development of codes across species clearly involves beings with radically different multimodal sensorial abilities and relevant aspects of the world; in short, the particular semiotic world of an organism which the founders of biosemiotics termed its *Umwelt* (plural *Umwelten*; henceforth here *umwelt/umwelten*).[106] In addressing interspecies communication we must therefore attend to beings who can perceive through some modes but express back in others: a horse who picks up a shout cannot shout back, but might flatten its ears and roll its eyes in response, for instance. We might speak of 'translation' across modes, or indeed, consider how concepts emerge at the integration of modes. This capacity, however, is actually perfectly usual in the integrated, multimodal perception and communication common to members of the same species. Human communities often conceptualise and communicate across modes; in sharing lunch, for example, a gesture might lead to the passing of a dish to someone who responds with verbal thanks, or a smile. We combine the language of speech with other sensory modes too in sign language, talking drums and, indeed, writing. Perceiving at the integration of modes is thus the norm for people, and consequently can be extended to interspecies communication in wider naturekind.

Communication might surely be more effective amongst beings who share the same sensory repertoire, yet sensory and *umwelt* crossover permits communication across species, however imperfect. Sensory alignment can also improve through living together, and can be worked on, in efforts to develop a frame of communication. Thus reindeer herders learn to listen to and pick up on numerous reindeer signs so they and their reindeer can move safely, harmoniously and efficiently together. People working with a dog or a horse codevelop a shared language of sound, touch and movement, and might refine this into the highly accurate language required to perform in dog agility competitions or dance together in equestrian dressage. As we learn to pick up the grammars and signing of a horse, so the horse learns to pick up the grammars and fields of signification of the rider. Technologies extend capabilities: the dog trainer's whistle pitched too high for human hearing; the horse trainer's repertoire of lunge and long reins, bits and nosebands, to assist the codevelopment of shared signals of touch and movement; or indeed the digital and AI technologies that are now enabling people to discern the songs and syntax of whales and birds, and to communicate back to them. So whilst communication across species is unlikely ever to be particularly accurate without reasonable sensory crossover, that does not mean that codevelopment of communicative 'languages' cannot be achieved.

Much looser forms of communication across species might still follow linguistic principles. Paul Greenough describes how, for example, many members of the Gujar community living in the Sariska Tiger Reserve in India have devised a protocol for how to meet and greet a tiger, with specific gestures and vocalisations, 'that has proved remarkably successful in terms of minimizing human casualties'.[107] Greenough regards this as an interspecies accommodation of a sort that most species must make with most others most of the time. We should not interpret the shouted warnings involved as a directly spontaneous reaction, or perhaps a prelinguistic one, because this would be to overlook how they acquire meaning according to structural principles, as part of a structural field of meaning and through experience in a particular linguistic community of people and tigers. The ill-fated young American Timothy Treadwell managed to develop modes of communication with the particular grizzly bears that tolerated his presence over the thirteen years he lived his summers amongst them in an Alaskan park. He and those bears he came to know each presumably had their own framing or interpretation of what was going on and how to coexist—until, that is, he was killed by a visiting stranger grizzly unfamiliar with the communicative order that he had established with the others.

A question arises, more radically, as to whether the codevelopment of linguistic communities must be restricted to living beings. Can we not develop meaningful exchanges with our surroundings, whether we perceive them as alive or not? What humans perceive as 'alive' varies enormously across the world, and many peoples communicate with beings far beyond what Euro-American rationalists would construe as 'living'. As we explore in the final part of this chapter and throughout the book, across the pluriverse, our interlocutors include angels, demons, spirits and gods; rivers, seas and forests. These are 'beings' that variously command respect and that can deceive, and meaningful communication—and linguistic communities—are established with them. This is not because the people involved are in some way deluded. It is because an orderliness can be perceived in the signs and syntax involved; an experiential conversation established.

Emplaced Communication

Thus we establish orderly, regular exchanges of meaning with a place and the things in it: perhaps a particular patch of forest, sea or city neighbourhood; a place that we become attached to, familiar with, play with, read and protect.[108] We discern the emergent grammars of place—in its orderly seasonal changes, for instance—attending to the signs and syntax involved as flowers grow, bloom and die in sequences, as animals migrate and insects appear with the seasons. We thus establish a 'community of communication', with 'linguistic conventions', with the pulsating world that does not just involve organisms and life, but also the regularities in exchanges of meaning with specific places. Indeed for any particular group of organisms, a 'common language' inevitably emerges out of dialogue with place.[109] And place itself becomes integral to grammars and fields of meaning, so that a gesture in one place has meaning different from that found in another.[110] It is this that ties sensual experiences to familiar landscapes,[111] and to the common experience that emerges amongst beings 'dwelling' in the same place, as anthropologist Tim Ingold observes, where they 'can share in the same meanings that inhere in the *relations* between the dweller and the constituents of the dwelt-in world'.[112]

It is not hard to see how this is not one-way human 'interpretation' of place, but a language codeveloped through dialogue, for we can and do understand the assemblages of a place (or a spirit) as responding to us, just as we respond to it. Indeed, it is out of such dialogue that we come to know a place—whether in the changing soundscapes of a woodland as we enter, and the movements of birds or deer; in the changing growth and patterns of plants in places where we once made charcoal; or in the particular way dice or cowry shells fall as we ask questions of spirits in divination. Developing a common language is what makes a place or a spirit alive.

Organisms make and inscribe place as part of their territorial practices. This is a human thing, as people chip geometric insignia onto the rocks marking portals to parallel worlds, paint animals in caves, spray graffiti on walls, write books that end up in shops or on shelves, or send social media posts to digital networks. In all these actions, people inscribe meaning out there for others to interpret, and such inscriptions become part of a material, sensorial world. Yet people are not alone in inscribing landscapes with their signs. All beings 'make places' as well as live in them, with their huge variety of meaningful marks, scents and sounds. Thus ants leave trails of micromolide wherever they walk that others may follow before it evaporates. By such a simple trick, this social insect develops collectively the most efficient way to get from A to B—a trick so sophisticated that we humans have only recently learned to mimic it in algorithms to plot the course of parcel deliveries or interplanetary expeditions. Humans inscribe landscapes to help each other, and so do all those that humans live among. It is sometimes said that people read landscapes as they read books, but the two activities are essentially one and the same. In this sense all life 'writes' and 'reads', through multisensory marks, grammars and fields of signification.

Structural biosemiotics is at once a theory of perception and of communication, and so whether one simply perceives a landscape, a forest, sea or rock, or envisages a forest as having 'mind', a sea as having agency, or a rock as having sentience, is by the by. Answers will vary according to what people make of such things within the particular conceptualisations shaped within particular human social worlds (and, indeed, within more-than-human social worlds). Forestry sciences have codeveloped their language (through experimentation) with forests that their mechanomorphic metaphors render as mindless, but the 'Forest People' of Ituri in the Democratic Republic of Congo have codeveloped very different languages with their familiar world, which they render mindful, as a spirit. Whatever the case, however, it is easier to perceive as mindless a forest or a carved stone that we have never encountered before, and which exists outside any codeveloped language (and therein lie the dangers to our planet of coloniality). What is important is that in our encounters with forests and seas we grapple to learn their signs in mutually constituted fields of meaning, and to discern significance from their syntax. A structural-biosemiotic approach thus displaces the focus of communication from any 'mind' and its essentialisation. It does not require a knowing cognisant self to 'lie behind' the communicative messaging of our 'interlocutors'. It sidesteps distinctions between the human, the non-human and the non-living that are rooted in authoritative philosophies of mind.[113] It sidesteps restrictive notions whereby some kind of dialogue of minds, interiorities or consciousness is necessary for symbolic semiosis. It renders irrelevant the assumptions and philosophical speculations regarding human uniqueness that have chronically been associated with debate on these matters. Although encompassing an interest in how fields of meaning are constructed across beings and collectivities of different sorts, it envisages no need to pin down such a 'mind'.[114] So whilst engagement in dialogue with rocks, rivers, winds or oceans might easily be dismissed as kookiness by those raised in scientific traditions, we might well ask what the conduct of natural science itself is, if not precisely such a dialogue, aiming to discern its own signs and regularities, its laws, its syntax. Physicists and chemists establish dialogue with the world around them, irrespective of whether they do or do not construe it as an interlocutor. The attribution of mind, whether to non-human beings or indeed between humans, is a matter of social construction. For how many times have humans not construed other humans as 'mindless'? Or machines as intelligent?[115]

We usually only become conscious of the signs and syntax that unconsciously shape perception and communication in the instances when something is 'out of place'. British anthropologist Mary Douglas coined this use of the term in the context of her interest in how the grammars we live by often only become visible—or perceived—when they are broken; when we make linguistic or social mistakes. She thus drew attention to the 'negative' aspects of grammar; to our propensity to recognise immediately when a word (or thing, or deed) is out of place. Things 'out of place' somehow get noticed immediately, yet for things to be out of place, there must be a sense of order. To return to our meal example, the appearance of porridge at lunch is out of place, revealing the ordering that links porridge with breakfast. Sharing lunch on a rug contravenes the usual ordering of sitting round a table—yet in a certain location becomes 'having a picnic', experienced joyfully in its difference from the usual. This is a prosaic example, but the stakes are often much higher. As the title of her structuralist analysis *Purity and Danger* indicates,[116] Douglas revealed how what is structurally or conceptually out of place is often immediately perceived as 'pollution' or 'danger', just as what is perceptually 'in place' remains subliminal. An important feature of the structural basis to perception is thus that it not only allows for communication to be established through forms of orderliness, but also (and for the very same reason) enables danger to be perceived as instantaneously as any grammatical mistake made in narrow human language. Douglas, like most others in the structural tradition, did not extend her analysis beyond the human, but it should be so extended, as its evolutionary significance and logic is obvious even though it has been overlooked until now: it is of danger (the things 'out of place') that beings become immediately conscious, rather than the structural order against which the 'out of place' is identified, and which remains subliminal: they simply do not register it. Just as we can understand afresh the disruptive significance of Marcel Duchamp's 1917 urinal being out of place in an art gallery, as it revealed the entire structural order and edifice that had become 'Art', so we can understand afresh the consternation of badgers who see or scent us authors as out of place when we camp in their field at night. Structural biosemiotics thus not only helps us towards a more-than-human theory of perception, communication and aesthetics, but is also crucial to the immediate sensing of danger, and is thus a matter of evolutionary fitness. For this reason, the capacity to exploit the structural basis for language and perception should surely be numbered among the traits most selected for across naturekind.

A sense of being out of place conceptually, aesthetically or socially extends into a sense of being out of place geographically; and we know when others are out of place, too. People know almost instantly when things are out of place in bedrooms, offices, fields and art galleries. Workers rarely wander into the boardroom. Urinals are rarely exhibited as works of art, and when they are, noticing they are out of place is key to their significance. It is not that the subliminal grammar of language, narrowly conceived, is 'similar' to grammars of social order and of place, but rather that they are in fact part of the same communicative order. Such spatial dimensions to meaning often involve actual social and geographical location, of the sort written into socio-legal regimes of human ownership and the sensed anxiety of trespass that keeps us on footpaths and in our place. It is the sort written, too, into the aural-territorial regimes of birds and many other creatures.

Packaged Meanings

There is a further implication of structural analysis that is also of evolutionary significance. Simple signs can come to signify huge content, if they come to acquire what we might term 'packaged' meaning (as 'metasigns', in the parlance of semiotics). For instance, if I choose to pack a coat in a bag, it might be the particular coat that my lover gave to me that speaks of that particular birthday, of our history and of much, much more, so for us, in our small linguistic community of coupledom, if I pick out that coat (as opposed to other coats that it is not) it carries enormous 'sentimental' meaning. It has become incorporated into our private field of signs relating to coats that is rooted in our shared memory and experience. In this way it is possible for a small image or gesture, a small singular sound, to convey vast information. Such packaged metasigns, with their sense rooted in historical experience and learning, in turn make sense in relation to other metasigns: other packages that share the field of meaning. In the field of academia, we need only drop a name and reference to evoke an entire argument or analytical tradition—'Marx 1848'—an argument that makes sense in relation to other entire arguments—'du Bois 1903', for instance. That packages make sense in relation to other packages with no need for analytic decomposition further amplifies the conveying of huge amounts of information with amazing efficiency. The tiniest gesture, a glance to the sky, a slight shrug, almost unnoticed, can be so significant to those 'in the know'. It is around such packages that human social lives and cultures have emerged in all their complexity. It is what makes for 'in-groups'. Might such packaging operate beyond the human too?

That communication is thus inevitably entangled with relationships and their unfolding was at the heart of the work of Gregory Bateson, who from the 1950s was arguing presciently, regarding communication across species, that the message conveyed by any sign acquires meaning 'only by virtue of context'.[117] For example, a cat's mew when seeking to be fed can only convey this in the context of its established relationship with its carer, which is itself affirmed by that very mew. Bateson also treated this context as a higher level of message—a 'metamessage' that classifies the message—but observed, importantly, that the message and context are inseparable. Metamessages are about wider relationships between communicants, and all communication in turn depends on, and is always also generative of, relationships.[118] In this respect, there are similarities between Bateson's metamessage and structural ideas of the metasign, as both root communication in relationships and their historical conditioning. Yet Bateson did not explore explicitly or systematically how metamessages acquire significance in structural opposition to other metamessages.[119]

In structural semiotics it is not just signs that become packaged as metasigns. Syntax, too, becomes packaged as metasyntax, or meta-grammar. Structural semioticians point to various ways in which sounds become packaged into the composites we call words, then figured into sentences, but further, to how such sentences in turn acquire meaning within particular narrative genres or styles, such as the thriller, the stream of consciousness novel, or poetry, each genre making sense in relation to the others. Such packaging extends to pictures, exemplified in the conventions of painting genre or poster genre—the latter, for instance, displaying packaged conventions embodied in juxtapositioning of what is placed at the top and bottom, in the foreground and background, left and right, flatly or in perspective and so on.[120] This metasyntax is perceived only subliminally by those already 'in the know', attuned to the language of posters—but becomes obvious when the juxtapositions seem out of place. Equally, people interpret cinema through codes of packaged narratives: through the usual sequences in stories and films in which the orderliness of sequential relationships, perhaps combined with colour or music, embody packaged subliminal conventions. The packing of a bag, as we noted earlier, signals impending travel, but with bright tones and tunes could be part of a holiday movie genre, or, combined with sombre light and minor key sounds, part of a much sadder departure story in a tragedy genre: the bag-packing carries quite different meanings in each case. Is the layering of packaged conventions into narratives and genres something that happens just in the human communicative worlds of literature, art and media? Or is it also relevant in non-human worlds? The meaning of birdsong during the dawn chorus in the context of a multispecies sound medley is perhaps different from its meaning in individuated song, in a territorial or mating genre. A dog's combination of barking, leaping and tail movements might carry one set of meanings in a narrative of play, perhaps with familiar dogs or human companions, but quite another in a narrative of aggression against intruders.

A corollary, of course, is that efficient communication through packaged meanings, and the establishment of metasigns and the metasyntax of narratives, must be understood as developing within the run of time, through use and history. Indeed, this is learning itself, refining language in its widest sense together with cohabitants and with place.[121]

Those applying structural semiotic analysis to human worlds have been attentive to the ways in which social orders and power relations also become embedded in metasigns and narratives, and thus go unperceived, naturalising social order, hierarchy and power. Because packaged meanings conflate, they also obfuscate contradictions that they might come to embody. The purr of a cat, the wag of a dog's tail and the chatter of an African Grey parrot might thus become packaged, as far as their owners are concerned, as innocent, playful signs constructed in fields of meaning associated with forms of pleasure; but the forceful separations of these same animals from their worlds of origin for commercial reasons are thus obscured. In the act of conflation, the historical, contingent and power-laden nature of relations can be obscured or erased[122]— a possibility of which we must remain strongly aware as we proceed to take a structural-biosemiotic approach to communication beyond the human.

This link between narrow language and wider culture has long been intuited. When the Moghul emperor Akbar the Great deprived some children of all linguistic influence in his endeavour to identify the God-given language, the motive was actually to identify the God-given way to live and be social. Akbar had resolved to follow the laws and customs of the country whose language the newborns would speak after twelve years of linguistic isolation. A Jesuit missionary, Hieronymus Xavier, discussed this wider aim with him, anticipating that the experiment would help convert Akbar's empire to Christianity. But as the missionary wrote, some in the cosmopolitan court of the Emperor hoped

> that they would speak the Hebrew language; others that they would not speak anything but Chaldean; while the Hindu philosophers and mathematicians asserted that they must infallibly speak the Sanskrit language, which is their Latin. However, the twelve years having passed, they produced the twelve children before the king. Interpreters for the various languages were called in to help. Each one put questions to the children, and they answered just nothing at all. On the contrary, they were timid, frightened, and fearful, and such they continued to be for the rest of their lives.

On finding no God-given language, the Jesuit missionary reported, Akbar thenceforth 'allowed no law but his own'.[123]

Some linguists question whether linguistic theory can be extended to multimedia or to social/cultural analysis, and so by extension to naturekind, as we are proposing. Noam Chomsky has argued that human language is of a very different order, qualitatively distinct from wider exchanges of meaning in social and cultural practices; that it constitutes a separate system responding to a genetically programmed 'universal grammar', with its own evolution—a phenomenon that is distinctly human.[124] This has also been the position of many biosemioticians. Yet, as we now find evidence of pervasive communication through signs and syntax, and of social orders (cultures), across naturekind, the debate must be expanded. Language, reconceived as part of a wider semiotics in multimodal communication and social orders, can no longer serve as a basis on which we as humans can distinguish ourselves from naturekind; on the contrary, it is one upon which we and the rest of naturekind stand integrated. And even if narrow 'language'—speech and writing—is somehow distinct (as Chomsky would like to argue), the study of communication cannot be restricted to a focus on this alone, given that so much of our symbolic communication is not expressed through it. Either way, our communication as part of naturekind can be studied in the wider frame of structural semiotics; and either way, we must cast aside the assumption of human exceptionalism. As we open the study of communication beyond the human, so we must expand the scope of what it means to be social. It is the end of social 'anthropo'-logy, for there has never been a separate anthropo-social world.

Structural Biosemiotics Across the Pluriverse

In developing a structural-biosemiotic framework rooted in advances in the biological sciences and debates across (bio)semiotics, the social sciences and the humanities, we have been drawing on and contributing to a genealogy of 'truths' about communication that is firmly rooted in formal, post-Enlightenment science. For many around the pluriverse, however, social worlds have always been more-than-human, and that people share meanings with all sorts of beings and parts of the landscape has long been perfectly obvious and appropriate. It is as much a part of the lives and worldviews of many First Nation and Indigenous peoples as is to a cat- or dog-keeper's affectionate relationship with their pet, or the magic for an urbanite of a wander in a forest or a swim in the sea. As Koyukon people in Alaska describe, 'a person moving through nature is never truly alone. The surroundings are aware, sensate, personified. They feel. They can be offended. And they must, at every moment, be treated with the proper respect.'[125] Those who engage in such relationships and encounters beyond the human 'get' that they are in some way communicative, even if dominant sciences, education and popular culture do not provide ways to comprehend or express this—indeed, label it 'romantic', 'anthropomorphic', embarrassing or just plain kooky.

One might therefore question whether our elaboration of a structural-biosemiotic approach is overly rooted in formal scientific traditions; whether, in fact, it is too universalistic to be applicable across the pluriverse. Can it possibly help in comprehending such a diversity of communicative experience? Yet if, as we are arguing, communication across naturekind is subject to structural principles at a most basic level, this can help us understand more-than-human linguistic communities, languages and cultures in all their plurality, and how, as these principles manifest, they interact with more restrictively human historical experience and discursive formations that 'frame'—understand and represent in particular ways—what things and beings are relevant. Whether we humans construe our apian interlocutors to be a bee, a hive or the god of bees differs radically across human social worlds; no less than in our forest encounters, as regards whether we communicate with individual trees, tree species, an ecosystem or the forest spirits within. Different corners of the pluriverse provide radically diverse answers to the questions that arise, as do our varied sciences.[126] Is the earth alive? Is this river a person? For some peoples living in Amazonia, for example, animals, even trees, are akin to human persons in disguise, with fur, feathers, bark rather than human skin. They inhabit a single social and communicative world, in which all beings assume distinct physical forms.[127] Many who identify as Koyukon in north-west Alaska consider non-human animals to be constantly aware of what humans say and do, communicating amongst themselves in ways that are parallel to, though different from, human language.[128] In the Andes, mountains are often considered to be the embodiment of Pachamama, an ever-present deity who has the creative power to sustain or control life and fertility, as well as to destroy it through earthquakes. Features of the landscape—particular rocks, or the boles of certain trees—have status as shrines to her.[129] In Australian worlds, narratives relate the common origins of all beings in 'Dreamtime', when common ancestors emerged from the earth at identified sites; their routes and stopping places as they moved around can still be discerned in the material environment of rocks, water sources and woods. Before disappearing, each left behind some of the many beings existing today—humans, plants, animals—and bequeathed totemic affiliations between these, and with places. In Māori worlds, human beings, non-human animals, plants and elements of the landscape also share a common origin, history and future within a single genealogy, *whakapapa*; all are imbued with *mana*, a spiritual and physical authority and power, and *mauri*, a spiritual life essence.

Similar framings can be discerned not only around the pluriverse, but historically, in European pasts from which new analytical inspiration might be derived. Before the Enlightenment era, European societies frequently ascribed personhood to non-human animals before the law, putting pigs, dogs, rats and even grasshoppers and snails on trial in court.[130] A plague of caterpillars was once charged with trespass.[131] The French Renaissance essayist Montaigne encapsulated aspects of our argument when he wrote that 'there is [...] no rational likelihood that beasts are forced to do by natural inclination the selfsame things which we do by choice or ingenuity. From similar effects we should conclude that there are similar faculties.'[132] Modern philosophers are returning to the seventeenth-century reflections of Spinoza, who construed all things, not just those living, to be part of an eternal, indivisible, self-caused whole in which all bodies and forces continuously affect each other, ceaselessly generating new forms.[133] He inspires today's 'vital materialist' philosophers, who discern the world as an assemblage of lively, generative interactions and forces encompassing living and non-living objects, forces and flows.[134] More recently, James Lovelock's 'Gaia' hypothesis envisages that all substances, 'living' or otherwise, together form a synergistic, self-regulating complex that helps to maintain and perpetuate the conditions for life on the planet.

Such worldviews, or framings, provide varying answers to the question of whether rivers or rocks, for example, have character, personhood and rights. They contrast strongly with the biblical (and Adamic) creation myths that uphold notions of human exception from the rest of naturekind; they contrast also with Darwinian narratives which, even whilst they discern evolutionary continuity amongst all beings that are 'alive' (dividing them from the non-living), establish human exceptionalism inasmuch as they posit or imply the narrow notions of language and culture that we have seen reason to question above.

Yet extensions of language and sociality beyond the human are not to be found only among pluriversal 'others', modern or historic, but also persist more pervasively, however under-recognised or threatened they may be. A tour of the pluriverse could just as easily amount to a tour of anyone's own home town, where such diversity is also to be found. It would certainly encompass groups struggling for animal rights and ethics. We should even perhaps take a tour of ourselves, to see that, for all we might uphold established scientific framings of our relations with the non-human world, these are very unstable, and we frequently switch out of them. When we authors cuddle our cats, we switch away from our 'Western' institutionalised assumptions about the world and the institutionalised mechanomorphism that the established sciences encourage us to adopt. Who has not, at some time, considered an animal in human terms—experiencing thereby what has come to be dismissed by formal science and education as 'anthropomorphism'? Even amongst scientists, established framings are unstable: find a primatologist or bat specialist who has never developed an emotional attachment that might threaten their 'objectivity', for example; while scientists we know have, in private moments, contemplated ideas of transmigration of the soul across species—the metempsychosis pursued by Pythagoras. Looking across the pluriverse we find not only the capacity of people to understand and experience reality in the very different ways alluded to above, but also to 'switch framings': to lock into one set of understandings and experiences, and then to lock into another, even in the blink of an eye—to treat a rat one second as 'animal', and the next as 'companion'; or indeed to treat a lover one moment as companion, and the next as animal. We probe this in later chapters, exploring how we both frame and switch framings, and we consider the epiphany involved as we pick up on and experience such switches. An incapacity to switch from a 'dehumanising' (that is, denaturing) framing, that eliminates empathy with the more-than-human world, to one that produces it, is regarded in modern psychiatry as a sign of the human sociopath.

Attention to the everyday pluriverse gives us a wider canvas on which to trace how the principles of structural biosemiotics manifest in different contexts, and where different framings prevail, and thus how meaning-making and sharing, metasigns, narratives and genres—languages, in the wider sense in which we use the term—are codeveloped, contested, accommodated between, struggled over. It thus also tunes us into the politics of communication in naturekind; to the power entangled with meaning and the many ways such power manifests in political relations.

Companionships and Assemblages

Across the pluriverse, we find both more individuated communications that people codevelop with particular non-humans, whether for company, work, sport or the host of other everyday activities that bring us together, and communications amongst whole networks of multiple kinds of beings. The chapters that follow are broadly grouped according to these different focal concerns, which we term respectively 'companionships' and 'assemblages'.

Companionships are the focus of chapters 4–8. Donna Haraway used the term to describe the joint lives of animals and people bonded in 'significant otherness',[135] interested particularly in the dogs she lives with, loves, and competes with in agility competitions. Haraway emphasises that companionship is mutual, involving a two-way dependency, and that it often changes both human and animal partners; as each learns from and influences the other, humans and non-humans 'co-become' something different from the beings that existed before. Building on this notion of companionship, and the effusion of anthropological works which document the exchanges and emotions involved, we probe their communicative dimensions: how companionship is forged through, and shapes, conversations.

Obvious animal companions, such dogs, cats or horses, are the least of it. Companionships extend to birds nesting in dovecotes, caged or as avian windowsill visitors; to the fish in bubbling tanks, ponds or rivers. There are children whose hamster companions are housed in plastic cities, wheels and tubes, and others who make homes for their woodlouse companions in cardboard boxes. Companions impel our participation in a huge range of competitions: not just weaving through and running beside poles in dog agility contests or guiding sheep round hurdles in sheepdog trials, but also as we work together in equestrian dressage, jumping or four-time gallop on the race track, or race pigeons home. Sometimes companionship is entangled with religion—with monkeys cohabiting temples and cattle cohabiting streets. Companionship embraces, too, the potted plants that make our homes and the trees that make our landscapes, all contributing to making us who we are; and even the sourdough and its bacteria and yeasts nurtured in city kitchens and farmsteads alike.

Companionships are entangled with identity, then, although not in any necessarily singular, exclusionary way: 'tree people' may also be 'fish people'; we, the authors, are cat people, and also horse people, chicken people and houseplant people (as well as academics, parents and more). Everyone has multiple identities and subjectivities. The point is that some, often very important, dimensions are codeveloped with our non-human companions and the (often small, even dyadic) linguistic communities we form.[136]

Companionships are frequently entangled with making a living. Cattle people follow the rains and fodder of their range in African savannas; just as reindeer people do in colder climates. There are elephant people: mahouts cajoling strength and cooperation into companionships that move timber and stone, or that bring power to weddings and other ceremonies. Dairy farmers' companionship with particular cows is entangled with keeping herds well and productive, while for the homeless in urban streets, dog companionship may equate quite literally to survival. The economic aspects of such companionships are hard to separate out from other qualities—whether of reassurance, pleasure, solace, prowess, expertise—and the sense of reciprocity involved. Companionship is shaped too by how we frame the world we cohabit, so just as there are bee people around the pluriverse, tending hives from treetops to rooftops, from rural Kenya to urban New York, not all are harvesting honey on the authority of the god of bees, as do Philippine montagnards.

The array of companionships is endless, and all involved find themselves in some kind of conversation with their companions. But what is the character of this communication? We address this through our structural-biosemiotic framework, by choosing to focus upon five kinds of companionship, for reasons that will become clear: those between humans and, respectively, chickens, horses, plants, bees and bats.

As we begin to see when considering bees and bats, communicative connections are also forged amongst assemblages of multiple animals, plants, other living organisms and their surroundings. As people immerse themselves in these they both 'hack into' existing linguistic communities amongst non-human beings, and forge new human–non-human ones. Whilst the biological and ecological sciences might conceive of such assemblages as 'ecosystems' or 'networks',[137] and the social sciences use network concepts too,[138] we prefer the term 'assemblage',[139] which is less mechanistic, referring to looser, more fluid gatherings and connectivities amongst beings and things. Here we are interested in particular in their communicative dimensions.

The array of assemblages in and through which people forge communicative connectivities is necessarily infinite, as are the reasons they might do so. Living in and with the shaded worlds of dense trees, undergrowth and overstorey are forest people and the many forest creatures with which human livelihoods and subjectivities are inextricably entangled. Then there are grassland people, living in and moving across plains, downlands and savannas as herders, farmers or hikers, or gardening with their soils. There are sea people, with everyday lives entwined with fishing, swimming, diving, boating and floating in and on watery worlds and their aquatic elements. Many humans—perhaps most—are now predominantly town or city people, living in 'urban jungles' of buildings, roads and amongst a sometimes surprising array of plant and animal life, but on occasion making forays into the more obviously green and watery worlds for pleasure, leisure or from need, or finding them in some form within the city itself. Forests, soils, seas and cities certainly do not encompass the whole array, but they provide our chosen foci in chapters 9–12, again probing communication beyond the human through our structural-biosemiotic framework and through examples from across the pluriverse.

The entangled lives we explore in such assemblages are often tough, bound up in livelihoods that constitute a struggle. And yet there are pleasures and fulfilments too. In this latter set of chapters we draw out how the 'projects' and purposes of human and non-human beings in such assemblages interrelate; how aligning them better helps with getting along, and how more-than-human communication in a shared linguistic community facilitates this. For those who visit less familiar assemblages, venturing to parks, woods, rivers and seas, whether to fulfil a desire to 'reconnect with nature', to seek health or solace, or simply for the pleasure to be found there, we consider how these encounters are communicative, and how this shapes their effects and affect.

This second group of chapters addresses further themes in our exploration of communication in naturekind. One is the question of how these ephemeral assemblages can be said to communicate when they lack 'mind' or 'consciousness'. We will have encountered this conundrum in previous chapters, when considering companionships, especially with plants, bees and bats; but it now becomes more glaring. What of the material elements and forces—the winds, waters, chemicals—that are part of such assemblages? Can they, or indeed a whole forest, or a landscape or seascape, be said to share meanings, to 'have language'? As we will see, a structural-biosemiotic framework helps here, moving us beyond a narrow response to such a question and finding such assemblages indeed to be communicative. Another issue is the significance of time, space and scale: how structured communication is intertwined with the rhythms of days, seasons and years; with the juxtapositioning of small patches and places in larger land, sea and cityscapes; and with histories over decades or centuries. The significance of the packaging of meaning through metasigns and narratives will come to the fore in these explorations.

Finally, considering assemblages reveals—just as, but even more starkly than in the case of companionships—contrasts between the more-than-human, communicative worlds found around the pluriverse, and the powerful orders sanctioned by capitalist and socialist modernities alike, along with their supportive sciences that often seek to deny the existence of, or replace, those communicative worlds. Highlighting the contrasts involved, and exploring more connected ways of living, being and communicating in naturekind across the pluriverse, in turn paves the way for their greater recognition in politico-economic and environmental struggles.

Conclusion

This book extends insights from structural linguistics, semiotics and anthropology beyond the human. As we have argued in the present chapter, this provides a way to overcome the limitations of existing biosemiotics, and to create a rapprochement with the linguistic and cultural turn now evident in the biological sciences. It offers the latter an approach to understand how meanings are made and shared. It also suggests new ways to appreciate framings and experiences around parts of the pluriverse where the fact of more-than-human communication has long been taken for granted.

This chapter has thus built a fresh structural-biosemiotic paradigm for understanding communication across naturekind, according to which all beings are to be considered as communicating through multimodal signing processes, ordered structurally and subconsciously within grammars, mutually constituting fields of signs, and packaged metasigns and narrative genres. Shared codes build up over time, through experience and memory, in linguistic communities that transcend species, and indeed, transcend even the 'living'. Narrow human language, based on speech and writing, becomes a mere subset of this wider range of meaningful communication. In appreciating this, we reconfigure language in broader, more-than-human terms, and we reconfigure humankind as part of naturekind. This is the framework we now take out into the pluriverse, to explore how communication between people and a wide array of animals, plants and the assemblages they are part of unfolds; how meanings are shared; and how and why this matters.

3

Chickens

Chickens, from their origins in South-East Asia, today have a massive and worldwide distribution—constituting 70 per cent of all bird biomass,[1] and are deeply entwined with human lives and activities.[2] Their social and economic significance extends from the provision of food and livelihoods, to their growing popularity doubling as pets and companions, to use in cock-fighting and the production of medicines. Our question here concerns the ways in which intertwined human–chicken lives and activities involve communication.

We focus here on human–chicken entanglements where livelihoods overlap with companionship in small-scale settings, including on our own smallholding in the UK and amongst rural communities in Guatemala, Ethiopia and China, where our framework reveals the structured, multimodal communication at work in everyday encounters, from encouraging chickens to roost to managing threats from predators or working with them in healing. Whilst there are different framings of what a chicken is (ontology), and whilst human–chicken relations in these different settings lead to differences in how structural principles manifest, we find stronger distinctions to lie between everyday, intimate interactions and those prevailing in systems of industrial production. Our descriptions might seem somewhat banal, but in this chapter we to want illustrate in a preliminary way how and why it is important to provide a linguistic interpretative framework for very mundane communicative experience.

Multimodal Chicken Communication

At one level, human–chicken communication might be construed as being about sound; about humans learning the 'innate' sound-based signs through which chickens communicate with each other, and tapping into or mimicking these. Indeed, along with a new proliferation of domestic chicken-keeping in industrialised European and North American settings—its boundaries with companionship blurred—websites to inform people about chicken 'language', so they can communicate well with them and codevelop affective relationships, have proliferated too. Let us begin by quoting substantially from one among the many, 'Roys Farm':[3]

> Chickens actually can't talk like humans, but they make several sounds for communicating [...]. You can learn most of this language or sounds simply by listening to and observing your flock. You can learn a lot about the ways of communicating with chickens if you spend some time with your chickens on a regular basis. Most people who are not familiar with chickens are often surprised to realize how intelligent, aware, sensitive and fun the chickens are, and how interesting and sweet it is communicating with them. However, here we are describing more about the ways of communicating with chickens. Listen to the sounds closely and notice what they are doing when they make a certain sound. Each specific sound has specific meaning. So listen to their sound closely, and notice what they are doing after making that specific sound.
> Generally mother hens will make a certain clucking instinctively to her chicks when she is out and about with them. When the mother hen makes a clucking sound, this means she is calling her babies, and she has found something interesting to eat, scratch at or play with. Some roosters will also often make similar sounds and the meaning is the same. After laying eggs, most of the hens generally make a loud calling noise. And many times other hens can also join in, and it can go on for a few minutes. Some people say it's a yell of relief, and others call it a signal of pride. Chickens can make a growling noise occasionally. Especially the hens make this sound when they are sitting on eggs and if someone disturbs them. It's actually a warning sound and may by followed by a peck or an attack. [... Roosters] make a soft 'perp-perp' sound for calling the hens over to a good supply of food. And the hens also make an almost similar sound for alerting their chicks to a food source. Squawking: chickens make this loud sound if you grab or scare them. Sometimes other chickens run when they hear the noise, and other times they are attracted (depending on the circumstances) [...]. Some hens seem to be humming when they are happy and contented.
> You can easily communicate with your chickens by listening their sounds closely and also by monitoring their activities after making a specific sound. You can sort of teach them to come near you when called if you are far away from your chickens. You can similarly cluck to the food call 'chick chick chick... chickies!'. Do this regularly and your chickens will learn gradually. Using higher tones while doing this will be good. Singing can be a good way for communicating with chickens. You can actually make any kind of clucking-ish sound like 'craaaaaaaaaaaw cruk cruk crawwwwwwwwwwww!'

You should use a quiet low tone of voice. After regular listening, you will be able to sort out the warning calls of your chickens. Go running to them after listening to the warning call to keep them out of danger. If you are able to keep them out of danger, then it will be very helpful for bonding with your chickens and they will be in love with you. ☺

You should make special sounds for special occasions. Chickens are very intelligent and they are capable of learning. And if you use one specific sound or call for one task constantly, then they will learn the sound gradually. If you really want your chickens to respond to a certain sound, then make it constantly for the same task. For example, use a specific sound while feeding them and make the same sound each time you feed them [. . .]. After making the same sound during feeding time for several days, your chickens will come running every time you call them (because they love to eat). The more time you spend with your chickens the more you will be able to learn the sounds they make, and the quicker you will become able to communicate with them. Chickens are lovely creatures and they generally enjoy human company in most circumstances. You will need vocal cords and patience for talking or communicating with chickens. God bless you!

We quote this advice at length because it captures the mutual learning that codevelops between people and their familiar chickens. Whilst it implies that human–chicken communication is all about sounds, and that these relate in part to instinct (desire for food, or fear), the observations on habits and patterns in the ways humans get on with their chickens, such as the rhythms of daily visiting and feeding, also begin to point to more multisensory signs, and codevelopment in their more grammatical ordering.

Research emerging from the biological revolution finds that chickens have at least twenty-four distinct vocalisations, as well as a large repertoire of visual displays, and that these signs are deployed referentially in ways that are symbolic. Chickens reason by deduction and discriminate between numbers, and quantities, and in ordinal ways. They can remember where things have been, and can anticipate future events. Many aspects of chicken behaviour and communication are not attributes of a species or particular breeds, but matters of more individual and group character, shaped by social learning, as hens pass on individual behaviours to chicks and all chickens watch and learn from each other. Their communication extends to abilities to outmanoeuvre and deceive one another, and relates to social hierarchies and 'pecking orders' they establish within particular groups—or cultures.[4]

Considering human–chicken communication through a structural-biosemiotic framework builds on these insights but focuses on how the multimodal signs in chicken communication make sense in relation to each other, and thus within wider fields of meaning and syntax that relate to situated experience. It suggests that we should regard all signs, such as the cluck of laying an egg, or calling the chicks, as gaining their significance (to other chickens, and to humans) as part of a wider field of signs: as biological anchor points in such fields, perhaps, but crucially within fields of signs that acquire meaning through their grammatical ordering and emplacement.

Structural Biosemiosis in Everyday Human–Chicken Life

To explore these themes more deeply, we can start with a common situation at our Sussex smallholding where we have kept a small group of chickens for the last few years, for eggs and to enhance our lives. It is early evening and they are wandering about the grassy yard where we let them out a while ago; they are scratching for grain and insects, nestling in piles of hay, moving around the places they know well, not straying too far. In these habits of movement they are interacting with each other, often moving as a group or at least keeping track of each other and never being too distant. We now want to bring them into their coop, so that when night falls they will be safe from the fox that our human knowledge tells us lives in the next field and will otherwise kill them. From familiarity we know the places in the yard where they are likely to be, and come round behind them with the practised technique of waving our arms. We say 'blub blub blub', in a way that resembles their sound signs, but is also the particular sound we have developed to communicate with them for this purpose. The waving and the sound operate together, as combined signs. And they react to the waving and the sound, starting to move away from us. There is a grammar to this communication, and it is structural; the signs (sounds/gestures) carry meaning in relation to each other in an ordered way, established through repeated practice that is emplaced and temporal. They are not our signs for 'get out of the barn', which involve sharper vocal signs, and other gestures of frustration, and they are not signs related to feeding, which involve the appearance of Tupperware boxes, usually in the morning.

Then we pick up that they are not very keen on returning to their coop, as they have not been out long. We read this from multiple signs: they run away from us as opposed to walking calmly; they ruffle their feathers and bounce in a disorderly way, rather than their more compliant gait, and they seek to take routes that are different from the path across the yard towards their coop. So they are both resisting us and signalling their resistance, through patterns in their head movements, their leg and wing movements and their directions, that convey something to us and also to each other. These signs of resistance make sense to us in relation to the same overarching grammar that ties meaning to pathways and space, sound, gesture and even the passage of time. We then start to get more insistent, reflecting the fox danger that is in our minds, even if not known by the chickens. So we clamber over the gate and root them out of the sheep shelter. We turn to a second long-practised technique—the waving of sticks. We get two bamboo sticks and hold them out as we walk behind the chickens. The sticks introduce a new sign into the field of meaning, the absence of which contributed to the chickens' earlier rebellious reaction—as every sign always carries the meanings of what it is not. We use the sticks held wide apart in a static, calm position to convey direction. This is distinct from a more vigorous waving or banging of the sticks on the fence or ground, which the chickens have also experienced. Structurally, there are possibilities for the sticks to be there or not there, still or wavy, quiet or banging. In this case, the chickens react to the calm directional sign, and move towards their coop. But if they had not, we could have used the sticks in a wavy or banging way, and the chickens might have responded. One might wonder whether these persuasion tactics are collaborative, or coercive—to make chickens follow human desires and purposes (that are for their own good and safety, though they do not know that). But they are certainly communicative, which is our concern here. Raising the question reminds us that communication is not without power, and one might think of a 'discourse of the waving sticks'. They key point here, and the reason for describing at length what might seem a trivial situation, is that the chickens and ourselves have communicated through a set of multimodal signs: the chickens signing to us, and us signing to the chickens, the meaning rendered legible through an overarching grammar established and sedimented through repeated practice, such as to become a set of subliminal grammatical rules. Vocalisation is but a minor part of this language.

Our chickens have, in the yard, paddock and barns that they wander in, a habitual set of places; a territory. They do not just move around in this, but inscribe it: scratching up areas of the grass, creating dents in piles of hay in the corners of the barns where they nestle, and laying eggs in habitual places in their nesting box or corners. But places and habits can be disrupted. Occasionally one of our chickens strays off its territory; it squeezes through a gap in the wire on our yard gate and wanders across the road to the public green on the other side. To surprised neighbours, the chicken becomes matter out of place, triggering attempts to chase it back over, and re-normalise a chicken-free green. The chicken itself is disoriented—as signalled by its flapping wings and somewhat directionless running around. Neighbours struggle to persuade the chicken back across the road, as they do not know or use the habitual arm movements, sounds or even sticks that the chicken has become used to. Disruptions also happen in the chickens' habitual places—such as when the tractor passes their coop, throwing up stones, or someone drives a vehicle over the grass yard where they are wandering. They signal their fluster by flapping their wings, clucking and running in a disordered way. These examples of matter and things out of place, to which the chickens and humans respond, reinforce the significance of the overarching grammar.

There are time dimensions to this communicative practice, interacting with space and place. There is a usual length of time for which we let the chickens out, and an order to this, corresponding roughly to the time it takes us to do our other farm chores and activities. If they have been out for only a short time they are a bother to get in, and, aware of this, we often avoid letting them out if we are just making a rapid farm visit, thus cementing the habitual time pattern. If they are still out in an evening and darkness begins to fall, they will often take this as a sign to enter their coop spontaneously to nest for the night. The neighbour who keeps chickens in the field next to ours uses a technology that raises and lowers the door of their coop automatically with sunrise and sunset. That is part of the grammatical structure for their particular group of chickens. This has become how things are: the rhythm of the door is part of the ordering of their emplaced world, and, were it to be disrupted, this would reveal that very ordering. Were the door not to open or shut as usual—for instance if its battery were to fail—this would amount to things being out of place and time. When the chickens in the unopened henhouse in broad daylight cluck and flap, might we infer that they are communicating not just wanting to be out, but being in disorder?

Zoologists, much like the advice-givers at Roy's Farm, might interpret all this chicken behaviour as linked to instinct: the desire for food, the fear of a loud vehicle or a stick. But our framework would suggest that, to the extent that these behaviours are indeed instinctual and biological, they nevertheless manifest in the communication of the chickens in our two yards very differently. This suggests that anything biological or instinctual is anchoring differently in wider, emplaced fields of signs; in short, in different established languages or dialects writ large. These are combinations of signs and their grammatical ordering codeveloped through relationships between different chicken guardians and their chickens. What chickens do depends on routinised communication; on structures codeveloped with humans, through interspecies communicative encounters and practices codeveloped through our interactions with these particular chickens, such that they have become habit—what sociologist Pierre Bourdieu once termed 'habitus'.[5] Where this habitus differs from the term as used by Bourdieu is that in this case it is extended to a more-than-human world.

While emphasising structure and habit, it would be impossible to understand how such human–chicken languages have developed without paying some attention to the agency of chickens. Sometimes chickens initiate communicative encounters. We find that our chickens sometimes follow us, or jump up and peck at us. This often happens when we are relaxed, sitting about and chatting in the yard. A conversation unfolds which sometimes seems to be partly about food—they signal that they want some, and we go and fetch grain or an apple, and feed it to them by hand; but appetite satisfied, it continues, with the chickens seeking further attention, and a more comfortable mutual 'being-with'. There are sometimes ambiguities at play; we cannot know exactly what our chickens want at a given moment, nor they us, but we find ways of accommodating and getting along. Reflecting on the idea of agency leads us to ambiguity rather than certainty. Some ambiguities concern how chicken agency is embedded in the chickens' sensorial and conceptual world that we cannot know, but which is referred to as their '*umwelt*' in the biosemiotic tradition. Others concern chicken sociality, and whether the agency is that of a particular chicken, or is more relational between chickens and perhaps even us as humans in our relaxed mode. The other ambiguity about agency is whether our conceptualisation of chicken agency is particular to our corner of the pluriverse and the framing of chicken ontology that it involves. These are all ambiguities, however, that relate not just to all multispecies communication, but to human–human communication too.

There are also moments when habitual conversation can be riffed on by either party, by communicating in a way that breaks the mould. Our chickens occasionally do this by starting to 'run amok' for no obvious reason. We might reflect that something is wrong—perhaps rats around; something new, anyway, even if we do not know what it is—but would respond by trying to restore order (calm the chickens down, get them in). We as humans occasionally riff on order, as we did when chasing the chickens off the tubs they habitually pecked in because we had newly planted flowers in them. A sharp shout and waving arms made an impact in this place and at this time precisely because it was out of the ordinary. In this case we might have wished the riff to lead to a revised grammar, in which the chickens would not peck the flowers—but we failed, and ended up having to re-plant the flowers in pots too high for the chickens to reach. Agency, and changes in the use of signs, can sometimes lead over time to shifts of emplaced grammars, but do not always do so.

These communicative interactions are not characteristic of all chickens. There is a particularity to them. At the most immediate level, this is the particularity of them being these chickens, with these humans, in this place (with communication patterns different, in minor ways, from those even of our close neighbours). They are part of what Lestel would call an interspecies, or hybrid, culture that has developed in the emplaced setting of our smallholding.[6] Language (combinations of signs and their grammatical ordering) partly reflects the status of these chickens as forms of life: 'domestic' chickens, and more precisely so-called 'hybrid layers' bred in Sussex—factors which must condition their particular *umwelt*. It is probable that communicative interactions and their sensory characteristics are linked to the physical inheritance of the chickens' breeding, and that structured grammars and biological anchors within these grammars have been nurtured through the process of domestication. The characteristics of both language and culture are thus dependent on the longer-term relations between chickens and their human breeders.

Framing Chicken Ontologies and Communication Across the Pluriverse

Such communicative grammars might operate a little differently in other parts of the pluriverse, where people frame the status of a chicken, what kind of a thing it is, differently. For instance, amongst Nuoso people, a Tibeto-Burman group living in the Liangshan mountains of China's Sichuan and Yunnan provinces, the rich repertoire of 'creation stories' divides animals and plants into twelve categories: six composed of bloodless plants, with the remaining six (the frog, snake, vulture and other magnificent birds, bear, monkey and human) composed of animals with blood. Chickens are defined outside of these categories, with a distinct creation story and role in the cosmos. As anthropologist Katherine Swancutt highlights, this gives the chicken significance as a sacrificial animal and in the use of its body parts, and especially its eggs, in divination.[7] The Nuoso chicken ontology underlies these birds' capacity to play this mediating role in what we, in this book, can consider as communicative practices implicating humans generally, specialists such as diviners, and various spiritual entities. At the same time, many Nuoso community members also frame chickens (and communicate with them) in more pragmatic ways, in everyday livelihoods and collection of their eggs, in which regard the sacrificial aspect of their status fades into the background and interactions seem far more similar to those we might encounter in, say, England. Indeed even in relation to sacrifices, there are more prosaic framings: one specialist concluded that the chicken is a popular choice in rituals because, being a common bird, it is inexpensive to slaughter and an easy source of eggs. Framings, and the communicative practices they shape, are thus not singular; people can switch between them—and not just people, one can presume, as signs and metasigns that chickens themselves pick up on can acquire different meanings in different locational and other contexts.

In some Mayan communities of Guatemala, where chickens live amongst people in the villages of the highland cloud forests, chickens are considered as selves in their own right, with capacities for reflection and a kind of reasoning, glossed in the Q'echi dialect as *na'leb*, the capacity to know something or someone, and to have intention. We frequently think this too of our own farmyard chickens. As the anthropologist Paul Kockelman relates, this underlies readings of intention, even trickery and deception on either side of the relationship between human and chicken. For example, he describes how 'a woman spent one morning watching her chicken rooting about in the underbrush, thinking that it had left a number of eggs there. She told me that it had a na'leb, in that it didn't want her to find its eggs. Similarly, when a chicken flew through a window into a home, the owner joked that it had a na'leb.'[8] The shouts and waving that such Mayan women use to bring their chickens in at night—here to avoid the predatory chickenhawk—seem very similar to our gestures in Sussex, albeit codeveloped and emplaced differently. However, these Mayan women interpret their conversations as being with highly rational beings, with whom one might try to negotiate, or even deliberately deceive, such as by moving ahead and giving the command to come to eat—a 'chik-chik-chik-chik-chiiik' sound—even when the women had nothing edible to offer. In these Mayan communities, women's lives are more entangled with the life and health of hens and their chicks than with any other animal. Women feel they influence chicks directly by their actions: for example, it is believed that if a woman sleeps in a foetal position, with her arms held around her head, the wings of her chicks will be twisted and unable to escape the egg. A woman's experience can thus overlap with that of her brooding chickens, with consequences for the newly hatched chicks; a relation that is felt to be similar to that with a pregnant woman's own unborn baby. This kind of unity of experience and the associated emotional ties are only experienced with babies and chickens, and captures some of the deep affect rooted in mimetic possibilities that characterise relations with chickens in these mountains. This raises the intensity of communicative practices, and of anxieties should care rooted in communication fail—and the chicks die, or the chickenhawk prevail.

One wonders, though, whether such framings are unique to Meso-America. Do aspects of them not also pervade the everyday lives of European chicken-keepers, especially in our entangled companion relations? We in Sussex (or the readers of Roy's Farm information) might also imagine our chickens as reasoning beings, and the sense of unity we feel with them—an affective, even loving bond—and the sadness and distress when a fox successfully attacks, are surely not so different from what the Mayan women feel in relation to their chickens and the hawks that prey on them.

In Ethiopia, in peri-urban Addis Ababa, Melanie Ramasawmy also found women expressing very strong relationships with their chickens, often declaring that they see and love individual chickens like their own children.[9] This was despite these being 'exotic' birds, supplied in batches of twenty-five to each woman as part of an African Chicken Genetic Gains project. As Ramasawmy puts it, 'even though these are different types of chickens than what they are used to, they are still attached to them'—an attachment (a connection) forged through the habitus of communication.

Whilst the different framings of chickens and human–chicken relations in the pluriverse might underlie some differences in grammatical structures and fields of meaning, then, framings are not singular anywhere, as people can switch between them. Strong continuities coexist with variations. The starkest distinctions lie not between scientific 'Western' framings and 'others', but between everyday interactions such as we have described, amongst humans and their relatively 'free range' chickens kept for pleasure and livelihood, on the one hand, and human–chicken interactions in systems of industrial production, on the other. Economic and social conditions geared to mass production, and spatial conditions involving incarceration (from crowded barns to battery cages), entail radically transformed structures of communication. Communication with humans in these circumstances is deeply impoverished, restricted to those chicken signs of distress that industrial farmers might choose to pick up on, ignore or harden themselves to. This severing of conversation is a feature of the broader severing of connectivity—a hardening of hearts—that has been noted as necessary for workers to cope in other settings where animals are raised and killed for human purposes. It is a separation that has been most elaborated in studies of architecture, worker practices, psychology and health in abattoirs.[10] To keep chickens in battery cages, people have to stop relating to them as communicative beings; to do otherwise becomes stressful and emotionally damaging. Physical separation and non-communication thus go hand in hand with conceptual separation and non-communicativeness, so enabling the particularly violent forms of animal exploitation that are practised under the conditions of mass production in which chickens are increasingly kept. Many of those around the world who keep chickens for livelihoods and coexist in companionship could not countenance such practices, given their experiences of connection rooted in communicative relations.

4

Horses

Whether for work or pleasure, sport or transport, human–horse relations are a recurrent feature of social life and history across the world and are always entangled with companionship of various kinds. The very intensity and diversity of equine relations thus make of these a helpful lens through which to consider how they are entwined with communication.[1] We will draw here on our own experience of living with and riding our companion horse on the hills where we live, as well as on examples from Europe and Central Asia, to consider how communication is refined in activities ranging from dressage and showjumping to mounted archery and long-distance travel. We tease out the significance of multimodal signing and grammars, however 'embodied' these feel in the intense connectivities and sense of co-becoming that horse and human establish, even to the extent of seeming, 'in the moment', a single being. Whilst this chimera is depicted in myths or ethnographies from around the pluriverse, it can also be sensed in the everyday experiences of many human–horse companionships, wherever they may be.

Equestrian Communication

To say that there is communication between a horse and a person is banal; all those who live with and ride horses know it, just as all dog-handlers know there is communication between human and dog. The challenge is to understand what is going on. For many ethologists, wielding their Occam's razor of simple, parsimonious explanation, such communication reduces to the inculcation of cue and response. Biologists currently recognise a widening range of multimodal cues, or signals, from people that horses pick up on, as well as those from horses to which handlers respond, in encounters in which both horses and humans have agency.[2] Biologists also reflect on the opportunities and the limits imposed by horses' and humans' different sensory abilities, and arenas of crossover. Certainly not all the sensory abilities through which horses communicate amongst themselves are easily available to humans.

Scent, for instance, is critical in inter-horse relations; people might try to mimic smell-based horse greetings—by blowing gently up a companion horse's nostrils, for instance—but with ambiguous outcomes.

Others conceive of human–horse communication not as mechanomorphic cue and response, but as an encounter of thoughts, intentions or 'minds', yet find themselves puzzled. As anthropologist Rosalie Jones McVey recounts of her discussions with riders in eastern England,

> The exact extent, and mechanisms, of horses' mind-and-body-reading capacities were mysterious to many riders. Could they really telepathically 'read' pictures from the rider's mind, as Georgie, a 58-year old well-respected riding instructor, suspected? Or was this idea too 'woo-woo', she wondered? Leslie, a 24-year-old who had recently completed an equine science degree, speculated that perhaps there was something special about their evolution as herd creatures that could explain horses' almost magical 'natural' capacities to connect deeply with others. The recognition that horses might pick up on a person's internal thoughts and moods is not only exciting and enlightening for riders, it is also uncanny, enigmatic, and even, sometimes, troubling.[3]

Anthropologists have responded to this trouble, and the desire to avoid anthropomorphism, let alone 'kookiness', by considering human–horse encounters as somehow 'embodied'—following lines of inquiry that emphasise how intersubjectivity, attunement, connection and empathy emerge from exchanges that are 'intercorporeal': in which minds and bodies are inseparable and permeable.[4] This approach envisages how communication takes place through direct sensory exchange, and the importance of this for feelings and emotion in human–horse companionship. Horses have thus become a focus for ethnographic and theoretical inquiry into embodied intersubjectivity in multispecies relations more generally, and into the phenomenology of direct, unmediated sensory exchange, conceiving of human–horse communication as somehow prelinguistic, or non-linguistic. Observing human–horse communication through a structural-biosemiotic lens does not lead to a negation of these insights about its embodied, emotive character within a theory of language beyond the human, but rather integrates them, for (as we have argued earlier) human language is itself just as embodied.

We can start personally, with an account from Melissa:

> I am riding Louis, our family's companion horse, in our field in the foothills of the South Downs in south-east England. I ask him to walk and then trot around a circle in the corner, squeezing his sides with my legs (not an absence of pressure, and not a sharp kick) and he responds by moving forward

in a regular rhythm. As we circle, I use coordinated pressure on the inside rein and outside leg (not the other way around) and he responds by bending his neck and body to the inside, moving comfortably around. Startled by a bird flying up from the hedge he pricks his ears (a sign in the wider range of ear positions), raises his head and swerves to the right; I respond immediately by moving my hands apart to alter the rein pressure, asking him to drop his head again, at the same time moving my weight slightly to the left and squeezing with my right leg, asking him to move his hind quarters back to the track. If I'd waited too long, if my timing was off, he might have continued the shy. As we reach the track, I straighten up. As we resume our pace and rhythm, but not before, I scratch his neck—a sign he has learned as a 'thank you'—and affirm vocally saying 'good boy'.

In this very brief conversation, meaning is created not through (prelinguistic) cues, but through signs that acquire sense within in a wider field of signs and their meanings: this kind of pressure, squeeze, shift or movement, not another. Signs—pressure, weight shifts—are coordinated and sequenced in a syntax that is essential to effective communication; getting them out of sequence would result in confusion. Here we have both signs within fields of meaning, the significance of syntax and sequence, and indeed the packaged meaning—the metasign—of the scratch. Whilst the horse responds to these human signs with changes of pace, position, carriage of head and so on, the rider responds to these in turn and to other signs from the horse, from pricked ears (not relaxed, floppy ones) to a tense (not soft) body, in an unfolding conversation that, as we go on to probe below, can almost produce the chimera of a single body.

What is going on here can be understood as a language, involving the coordination of multimodal signs and grammars within a mutually constituted code. Such basic structural principles of communication have codeveloped between horses and humans over long periods to permit riding. The established codes can be written down in disembodied, formalised training protocols and manuals, but an analysis must go further to look at the more nuanced codevelopment of communication—still within a structural-biosemiotic framing—that captures the far more intimate meaning-sharing that riders and their horses experience.

A classic example of a formalised depiction of equestrian language is to be found in *The Manual of Horsemanship*. First published in 1950, this is the official guide for young riders in the UK's 'Pony Club', but has become far more widely familiar in the equestrian world, its title being known and basic concepts followed in several countries. In this manual, signs are described as 'aids' that work together in a shared language that enables horse and rider to move well together:

The desired pace is created and maintained by using co-ordinated leg and hand aids. The energy created by the horse's hind legs and quarters is received by the rider's hands without resistance. This should give the feeling that the horse is going forward in balance, 'between the rider's legs and hands', forming a harmonious partnership.[5]

Lynda Birke and Kirrilly Thompson capture how riding is 'a strange form of communication. It is mediated by the sense of touch, yet there may be no unmediated hair to skin touch between horse and human. Rather, there is usually a saddle and bridle in between.'[6] As the *Manual* explains, such 'tack' that mediates sensory communication can take various forms, the type of bridle and bit altering the signing through the reins between the horse's mouth and the rider's hands—the 'contact'—and the type of saddle altering how the rider's leg aids and weight shifts are given and received. The advertising and sale of such equipment that facilitates equestrian 'language', as well as the provision of advice itself—from manuals to training programmes in clubs and riding schools—has become a veritable industry.

Whilst the *Manual of Horsemanship* relates the codes of established UK riding traditions with a genealogy rooted in the British cavalry, there are many other such codes across the world—essentially language variants, or regional dialects.[7] These range from so-called 'Western riding'—the stuff of today's ranching and herding cultures in the United States that developed from Spanish 'conquistador' traditions and the variants developed in First Nations traditions—to the even more hands-off techniques of Central Asian styles. The communicative signs in Western riding work slightly differently, with the horse moving away from pressure of the rein on the neck, for instance: thus 'neck-reining' usefully enables the rider to signal direction one-handedly, with the other hand free to wield rope or gun. New variants of equestrian codes, drawing together diverse traditions, are emerging and becoming formalised through training protocols such as those of what is coming to be called 'natural horsemanship'.[8]

The basic structural codes that enable equestrian communication are adapted and refined during training and practice in particular activities and disciplines, whether it be in the specialised realms of equestrian sports—from dressage, showjumping, eventing and its cross-country jumping to flat racing, steeplechasing, polo and mounted games—or in riding out in rural or urban terrain, or navigating landscapes whether for travel or pleasure. Horses and riders learn adapted combinations of signs suited to the purpose in question. Related but different grammars have codeveloped around driving, ploughing, log-hauling and other activities in which horses pull or drag vehicles or other objects. Suffice to say that there are also grammatical principles involved here, and these too work through packages of signs with some sensory crossover which is parallel (such as bits and mouth pressure, and voice) and some which is different, reflecting unmounted interactions. These codes need to be learned: a horse or human brought up to drive would need to retrain to ride, and vice versa.

Whilst the equestrian languages and dialects of particular regions or activities must be learned, they nevertheless have much in common. A rider can thus mount a horse that speaks a 'foreign' interspecies language and still ride, even though communication will initially be far from perfect, and full of confusion. Melissa relates, for example, experiencing this with horses in the Camargue area of southern France, who do not respond to attempts to take up an English-style rein contact in the way that an English horse might. In this respect, interspecies equestrian language is not so different from human spoken language; speakers have to learn and accommodate to each other's codes if they are to communicate.

Yet as we now go on to explore, equestrian language involves far more than the kind of formalised codes we can describe in training manuals. Rather, it emerges through the experiences of particular horses and people as they live and codevelop the structural principles of communication in specific companionships and in relation to the world around.

Personalised and Emplaced Communication

A structural-biosemiotic lens reveals how human–horse communication follows structural principles, but also how these are constantly refined—and in ways that are not only as subliminally embodied as the grammars and fields of meaning of human languages, but which are also emplaced.

Human–horse communication is thus a far more personalised matter, in which particular people and horses in particular contexts mutually refine communicative codes. This is how partnerships are built. Disciplines such as dressage or showjumping rely on it, to navigate dance-like sequences of movement or courses of obstacles, but everyday riders also seek and value close levels of mutual attunement with their equine companions, as Jones McVey recounts:

> One rider, Sarah, an insurance salesperson, told me with wide eyes about a remarkable experience with her horse, Silver, when she was able to tune into the profound level of responsiveness occurring between the two of them: '[After that lesson] I suddenly realized I am communicating with her the whole time. Whether or not I mean it, or whatever I think I am doing, I am actually communicating, because my body is communicating, and that is going to be influencing her, she is picking it all up [...] everything I'm feeling [...] so, even how I breathe, it is linked to her breathing (*long pause*). It was incredible.'

The process of endlessly attempting to improve communication, enhancing what the people involved might refer to as 'connection' or 'feel', is forever ongoing. Moments when 'deep connections' through communication are achieved interact with miscommunication. Riders often reflect on this among themselves, and it is a key reason why many seek riding lessons and further training. As Jones McVey observes, perfect attunement is ever elusive, and greater refinement through mutual learning leads the riders to be more critical of their own efforts, more aware of a perfection that remains beyond their grasp.[9] Something similar happens in partnerships between people, as through companionship we attune our communication styles to each other, and yet this ratchets up the expectations, exacerbating our frustration when we miscommunicate.

That equestrian communication 'has its moments', its ups and downs, underlines how it is not just codified, but practised and performed.[10] Thus Melissa and equine Louis became accustomed to the particular combinations of highly subtle signs that each responded to; and each would communicate slightly differently with a different companion. Yet to such partnerships and the conversations that ensue within them, horse and human bring their histories and previous experiences: legacies of past events—of a fright or a fall; of past gentle and empathetic companionship or violent treatment and inconsistent communication; and the lasting effects of these must also be negotiated in the conversations of the moment. The detail of human–horse communication is also shaped by the particular characters or styles that each partner brings to an encounter. Many phrases used by humans of horses capture such differences: thus in English a horse might be described as 'highly strung', or as having a 'calm temperament', or even as 'bombproof'. Sometimes these qualities are seen as breed characteristics which might be sought to suit particular human-defined needs: for instance, in the UK, contrasting the highly strung thoroughbred who dominates the racing and showjumping worlds with the calm, safe cob or with English native pony breeds regarded as being suitable for children to ride—though interpersonal variations often buck any breed trend. Successful companionship and the communication that underpins it involves mutual accommodation between the horse and people involved. As one of the continental European riders interviewed by anthropologists Anita Maurstad and Dona Lee Davis put it, this is not so different from how we relate to each other:

> I've also learned that if a horse has a personality, then you learn to work with that personality. You don't try to change the horse to work with your personality, because it's crazy. So I use the same when I deal with humans. You know, if someone has an angry personality you don't do things to develop the angry personality, you relate to them in ways that are going to be calm. You do that with horses, if you have a nervous horse, you don't want

to go in there all shaky and jittery and hyper, you want to go into their area kind of droopy and laid back.[11]

As in this example, characteristics reflected in people might lead us to ideas of horses having personalities or temperaments in much the same way as we do. Likewise, people also bring their own personalities to human–horse conversations, in ways that can shape present conversations and the companionship so forged. Thus Melissa's companionship and everyday communication with a very placid, gentle Norica cob, Jani, was something she found calming to her anxious tendencies, whereas her anxieties just wound up another horse with a nervy temperament, and the companionship did not last long. Yet as this experience and the quote above also suggest, these relations are not just a matter of character as something innate or biologically anchored. Nor need they be understood in terms of horse and human minds mysteriously influencing each other. Rather, they reflect linguistic communicative encounters in which the legacy of past (communicative) experience engages with the moment. This has been described as a honing of a 'responsive embodied channel of attentiveness'—an intercorporeality—between horse and humans. Vinciane Despret elaborates on how communication between horses and humans happens through small bodily movements, operating below the level of consciousness. This was the case when Clever Hans appeared to solve mathematical problems, and it happens too in riding: 'Unintentional movements of the rider occur [...] when the rider thinks about the movements the horse should perform. The horse feels them.' Despret focuses on instances when the horse's response reproduces the rider's movement ('the bottom of the rider's back makes a jerk which is the movement a horse will make to begin to canter, and so on'), arguing that 'attunement' is thus a matter of riders having learned to 'behave and move like horses'.[12] Yet human–horse communication is not just so directly embodied, or dependent on this kind of iconic sign; both in riding and handling it depends on coordinated signs and syntax. A structural-biosemiotic framework draws attention to how this works, as structural principles are adapted, and multimodal signs are coordinated in particular ways: the movement of a back combined with pressure and response between rider's leg and horse's side, or perhaps a click of the tongue, and a taut comportment conveying 'it's time to do something', in contradistinction to other comportments, movements and sounds and their sequences. People and horses learn to read these codes from each other as subliminally as any other language is learned. Likewise in handling, a calm tone, slack body and slow movements might relay calmness, but do so in relation to each other and in relation to other signs: to the sharp raised tone, a shout or a sudden push, perhaps, as are needed when a horse is poised to step on a toe; or to the horse's sudden movement and fidgeting, arching and snorting, as a signal that something is up. In this way signs and grammars—language—become established through experiences in which both horse and human are involved. Inasmuch as these communicative encounters 'in the moment' are performances, they are also structurally coordinated conversations; they are malleable, but not endlessly so.

In such particularised refinements of human–horse communication, place and emplaced experience become significant. Conversations unfold not just between a horse and a person, but in relation to signs in the world around. The simple vignette of Melissa and Louis that we started with would unfold differently if the place were not a field corner but a stony track descending a steep slope on the nearby South Downs; here the coordinated mutual signing in pressure and touch, weight adjustment and movement would be adapted to the terrain in subtle shifts that enabled the pair to get down the track safely, without slipping or tripping. In a similar way, those who plod behind the plough for eleven miles per acre and their horses are in constant communication not just with each other directly, but in negotiating mutual adjustments, in response to the subtleties of the terrain, that depend on and consolidate their partnership. Places create very different contexts for equestrian communication: from the open plains of horseback herders, to the deserts and mountains that people and horses co-navigate in travel, to the regulated and constructed spaces of riding arenas and sporting establishments. But these are more than just contexts, as Tim Ingold's notion of 'dwelling' captures.[13] Their features become signs in syntactical relation with signs that horse and rider might produce; they might even become integral to the horse's signing—as its foot dips into a hole and its gait alters, for instance, to which the rider must respond. Place thus becomes integral to the unfolding of more-than-human communicative engagements. These are multisensory—the touch and pressure of flat or bumpy terrain, hard turf or mud; the feel of wind, the sense of light—and yet they are also structural; coordinated to comprise a landscape in which human–horse communication takes place, and with which it engages.

The significance to communication of emplacement emerges not just in riding, but in the everyday care that human–horse companionships involve and through which they are forged. Horses living with people usually have a dwelling place, a territory, created, provided and shaped in line with human purposes, whether the small yards and fields of horses and ponies kept at home for human pleasure, the rows of stables and paddocks of larger commercial livery yards and racing, breeding, showjumping or dressage establishments, the corralled enclosures of ranches, or the open plains and commons of horses coinvolved in herding lifestyles. Spatial geographies are associated also with temporal routines: daily (some live in a stable, some in a field, some come in at night and go out by day, and so on); yearly (the stabled horse who has an outdoor holiday, the plains-grazer who is brought into a corral for the work season); or in relation to a horse's life-stage (the youngster who grazes freely until 'brought in' to human routines of work and learning; the retired horse 'put out to grass'.) Places and emplaced routines become integral to the grammars of companionship. Feeding might happen here and not there, grooming happens there, not here, or is carried out around riding and before feeding, and so on. Horses inscribe their places in many ways, rendering them familiar and fitting to their routines; some will always use a certain field corner to defecate, or will roll or lie down always in the same spot, the disturbed ground or bedding a sign of routinised occupation and habit (and a sign to the humans caring for them to respect this).

Emplaced communication also encompasses the spatial positioning of respective horse and human bodies, in relation to each other and to the parameters of the yard or terrain. Accustomed handlers often seek to maintain themselves and the horse at a mutually respectful distance, so that when the horse follows and moves as it might, it does not barge. Most horses are extremely sensitive to gestures, including those that relate to proximity and space. Many will convey signs of discomfort if approached too rapidly from the front, without time to sniff and see what or who is coming; and doubly so if approached suddenly from behind. Some speculate that this might be derived from long-established horse–predator relations, whereby those that attack horses in the wild usually come from behind, with claws. If so, such 'instinct' provides what we have been calling a biological 'anchor point' that is not simply instinctual, but forms part of spatial fields of meaning codeveloped in particular human–horse relations.

Whether proximity and bodily placing should actually be considered a human sense is now debated. Ethologists have recognised such 'proprioception' in horses and consider this as part of multimodal communication with humans. Certainly this aspect of communication—bodily closeness and proximity—is central to the emotional connections that humans and horses create. It is central to the claims of many specialists in human–horse communication, from those renowned as 'horse whisperers' (where whispering is more about gesture and proximity than voice) to those engaging in the practices that have come to be grouped and sometimes professionalised as 'natural horsemanship'.[14] Touch and proximity are also central to the many circumstances in which horses are used for human therapy and healing,[15] when some form of mutual connection through communication is surely of key importance. Underlying and enabling these specialisms is the more basic habitus of gesture and proximity as central to human–horse communication in general.

The significance of place and routine in communication with horses also renders significant their disruption, in 'matter out of place' or 'of time', highlighting also the importance of memory in shaping the emergence of communicative encounters. To cite another personal vignette, Melissa relates that

> Louis and I come upon a group of cows and their calves while on a Saturday morning ride on the South Downs. The cattle cluster around a gate and quite peacefully block the way. We are used to this route being either empty of livestock, or to finding sheep grazing there. I feel Louis tense his muscles, arch his neck and bounce a little, shaking his head; the signs I've grown to know as expressing that something is up. Louis has met cows many times before, but never in that place. Is it the surprise of finding them there? The crowd of them, their jostling and movement? A disrupted grammar of the usual quiet, still path—matter out of place? I chat to Louis inanely and we eventually walk through, keeping my own body slack and relaxed; trying to convey calm through my tone of voice and position. And so we go on our way. Later, back at our yard, I tie him in his usual place to wash him down and feed him, relieved to see that his elderly donkey companion is clearly visible in their field just the other side of the fence. Too often, we have arrived back like this and the donkey is nowhere to be seen, usually because he's tucked himself to stand in a sunny corner of their shelter. But on such days Louis will not settle until he's located him, bouncing and calling, and when turned out, will gallop round the field until he's found him. Attachment? Surely there is something of that. But over many years I've also come to read a matter out of place into Louis's concern; a disruption of the grammar of his place, which has a donkey in it. Part of his territory; even a marker of it.

That this description and reflection on these incidents is couched in terms of 'matter out of place' surely reflects our anthropologists' conceptual interests, but also a simply reasonable interpretation of human–horse experience. It is the fact of the usual emplaced, grammatical ordering that allows the shock of things out of place, or routines disturbed. As this vignette suggests, Louis shows signs of disturbance and upset when he cannot see his donkey companion wandering about in their shared field—a disruption to the emplaced order of things—and this becomes far worse if the donkey escapes through the fence into the sheep paddock, where he is still visible, but definitely 'out of place'. An odd object in the middle of the yard—a wheelbarrow left out, or the tractor parked not in its usual place, and Louis might react, too, signalling with sharply pricked up ears, rolling eyes, and a snort—signs that his human companions pick up visually as registering something disturbing. The same object in its usual place would pass unnoticed; just part of the usual emplaced grammar of the surroundings. The signs and syntax developed with place through such experience are not qualitatively different from the signs and syntax codeveloped between horse and rider in human–horse communication. Indeed they are integral to it. Horses that are so attentive to 'matter out of spatial place' are equally attentive to ill-judged signs out of place in the grammar of riding.

Horse and Human Co-Becoming and Affect

Being together in human–horse companionships is also becoming together. Both human and horse change, and become different through their ongoing communicative engagement. As Donna Haraway says of human–dog companionship and engagement together in agility competitions, 'partners do not precede their relating'. In a similar way, 'riders, as partners to the horse and vice versa, are relational categories arising from engagements in a range of intra-acting practices that form both riders and their horses'.[16] This co-becoming can be reflected in bodily changes. Practising and achieving physical and communicative connectedness changes the human body, as the rider builds core strength, balance and leg and arm muscles through applying the 'aids' to guide the horse, and the oft-ridden horse acquires the rounded, muscled hind quarters and neck ('topline') that signify not just fitness, but repeated working together. Rather in the way that dogs and owners are said to become lookalikes, there is a truth as well as a stereotyping in observations that each equestrian activity has its own bodily-adapted pairing: from the lean and wiry racehorse/jockey pair, to the cresty, rounded, balanced horse and equally balanced rider of dressage companionships.

Co-becoming is not merely physical, but also affective. In many settings, human–horse companionships affect the feelings and emotional and mental states of their human participants—something that is mobilised in creating therapeutic practices that involve horses. Whilst we would not seek to impute feelings to the horses involved, accounts do suggest a mutuality in this. Several of Jones McVey's informants spoke of cherishing their companionships with horses for their 'authenticity', and interpreted horse responses in their encounters as providing unbiased, honest feedback on their faults not only within the context of their companionship, but extending into wider life. Furthermore,

> receiving authentic feedback enabled glimpses of real self-understanding. There were depths of complexity to this understanding that could constantly be refined, such that riders could continually improve themselves as people as well as riders.[17]

Such co-becoming has hitherto been understood through ideas of embodiment and intercorporeality, but we have been arguing that there are structural-biosemiotic dimensions to the communications that contribute to it. Embodiment and intercorporeality are not distinct from structural biosemiotics; rather they are integral to each other.

Especially in riding, human–horse communications can become so well aligned that horse and rider merge to become a single unit, or at least a rider feels this. The agency in navigating a showjumping course or a steppe is the connected agency of a unit navigating or dwelling in relation to the course of jumps or the wide open space and its winds. Such connected agency depends on communication that is then hardly between a separate horse and rider, as individuals, but within a single entity, almost as a single body. Riders across many settings describe—and seek—such experiences of connection in which attuned communication is taken to the extreme, even if only for short periods. Thus European and US riders competing in equestrian sports speak of 'being in sync' with their horses as being central to both purpose and pleasure. As a dressage rider put it, 'I actually feel part of the animal, reacting to his body and my body. It's that connection that you start craving. Once you have it, you need more. It's almost an addiction.' Another emphasised riding 'mainly for the feeling for when you and your horse are in sync and everything that is communicated is fluid and it just, everything works out like, like you're one'.[18] Ann Game expresses it thus: 'When a horse moves freely, balanced, with cadence and lightness, it feels like floating and flying. Ecstatic. These moments of effortless airy floating, flying, so light yet requiring perfect placing of the hooves on the ground, so high, the wave-horse rising in-beneath me, are the moments we ride for.'[19] Riders and trainers value and search for experiences of 'true connection' between horse and rider.[20] Those who experience this often concur that it is not only humans who value this synchronised, interconnected behaviour: horses too take pleasure from it.[21]

At least within European and US framings, this can be understood as a fusing of human and horse 'minds' as well as bodies. At its best, the intercommunication seems so sublime as to synchronise minds, bodies and even souls into one being. This is the inspiration for the centaur metaphor and for the pervasive equestrian sentiment of horse and rider thinking, acting and being 'as one'.[22] Yet such experiences of riders in European settings (and with the class privilege that tends to be involved in enabling an enduring relationship with a horse) are echoed around the pluriverse, albeit described and represented within different framings. Natasha Fijn captures, for example, how riders in Mongolia, in herding and especially in the ancient practice of horseback archery, value the feeling they refer to as *khii mor'*, 'the vitality and heightened spirits one feels when the wind hits the body, and the sensory elation of flying through the air when galloping along, a feeling of being at one with the horse'. As one rider elaborated,

Modern day Mongolians talk a lot about *khii mor'* [. . .]. being with *khii mor'* means to be 'lucky', 'fortunate', in 'good health' and in 'good spirits' in general, also a fighting spirit to survive and a pride in your horse. That's what being happy is for Mongolians. *Khii mor'* is also related to the spirit of the horse. The person riding the horse has a certain feeling—it is not just for transport. Personally, I feel it when riding a horse, there's a happiness factor. You can talk about *khii mor'* and being in high spirits if you ride your horse and shoot your arrows from your horse and hit your target, that's when I feel *khii mor'*.[23]

Framings of the ontology of horses add to the ways human–horse communications are experienced. In Mongolia, horses are understood as embodying four elements of vitality, as represented in images of the 'wind horse' flying into the heavens, depicted on the flags that flutter in the wind on top of sacred cairns on mountaintops or high passes across the Mongolian Plateau. Horses carry a life energy connected with elements in the world around—wind and dust—and this can transfer to humans; the dust stirred up by racehorses bears good fortune, and after a horse race spectators rush over to the winning horse to wipe sweat from it with their hands, in the hope of transferring some energy from the horse to themselves.[24] Neighbouring Tuvan riders also speak of the life energy in the horse, whereby the spirit of the rider and the horse become interconnected.[25] Such notions of a 'spirit' in horses that derives from and connects them with the liveliness of elements in the world around is to be found in modern European settings too: take the many popular public displays with titles such as 'The Spirit of the Horse', or the representations in popular fiction or advertisements of horses galloping freely, moving with speed and power in ways seemingly entwined with the vital forces of land, sea and sky.

Such human framings depict horses as giving access to a wider communicative world that is alive with meaning beyond its physicality. Through establishing communicative identity, becoming one with a horse, a rider thus gains access to this lively communicative world. This is surely part of the feeling of being 'in sync': in sync not just with the horse itself, but with the wider energies it embodies and is interconnected with. Such a merging of self and identity may be temporary, even momentary; part of the experience of riding (as distinct from other parts of a human's life) and perhaps only of fleeting moments within that. But a switch of perspective does happen. One can pick up on a switch in the horse itself: the change of carriage that occurs during fluid communion, the signs of calm afterwards but also the resumption of a more usual way of being. There is something sensual, almost sexual, in this coupling, in the course of which one switches perspective from that of the individual to something more communing. Indeed this is not so different from human–human sex itself. It suggests that the sensuality, or sexuality, of riding is more about devolved agency, and what communicative connectivity feels like. And although one might find this depicted in pre-Enlightenment myths or ethnographies from around the pluriverse, it is also part of everyday more-than-human equine experience.

5

Plants

We turn now to consider the companionships that people establish with particular plants and trees—in their homes and in the landscapes they live in. It can seem 'kooky' to discuss communication with a plant, as even though science has revealed many ways plants communicate among themselves, these signing processes have little if any sensory crossover with humans and are easily construed as mechanistic. Yet structural biosemiotics attunes us to numerous ways in which people do establish communicative relations with particular plants that we can see as linguistic, and this underlies what we can reasonably call companionship.

Considering plants also sharpens questions concerning the varied ways in which people experience communication, and the thorny issue of imputing mind. As plants have no central nervous system it is usual to assume that, without consciousness, mind, or intent, communication cannot happen. As a focus on plants reinforces, however, communication does not depend on imputing a mind behind it. Much of the communication we pick up on from our human others involves gestures and subtle expressions that are unintentional, subliminal or subconscious—so acknowledges that communication can take place without mind. So it is with plants; but as we are arguing, such semiosis still needs to be understood linguistically, in structural-biosemiotic terms. A focus on plants will thus drive home that the making and sharing of meaning—and thus communication—is not dependent on the kinds of mind that science and philosophy have attributed exclusively to humans.[1]

Ethnographic examples of people's communication with houseplants, garden plants and with particular trees illustrate these arguments. These examples underscore the significance of emplacement and experience of a particular plant's life circumstances in establishing intimate communicative connections. We address the incompatibility between such place-rooted, accreted meanings of plants and trees, and the presumptions of transferability and replaceability in contemporary neoliberal approaches to plant and tree production, exchange and conservation.

Communicative Plants

The biological revolution is revealing a multitude of ways in which plants communicate with each other. Their many chemical signals are usually understood by biologists as mechanical exchanges of information, and by biosemioticians as semiotic processes. While plant biosemiotics, in common with biosemiotics more generally, focuses on so-called prelinguistic signing, considering chemical signals as indexical signs that convey meaning in ways that are direct, not linked to symbolic learning, biological research is also showing how combinations of signs that have been interpreted to have a syntactical quality convey meaning, and that plants 'have memory' that shapes responses to signals.[2] Suzanne Simard and Peter Wohlleben, whose popular writings use linguistic concepts to describe tree communication,[3] might once have been dismissed as anthropomorphic—yet the 'language' they describe is now proving more than metaphorical. Thus Wohlleben asks, 'Wouldn't it be interesting to know whether trees can [. . .] talk to each other?' 'But how?" he goes on to enquire. 'They definitely don't produce sounds, so there's nothing we can hear [. . .]. Trees, it turns out, have a completely different way of communicating: they use scent.' He suggests this is analogous to humans: just as '[s]cientists believe pheromones in sweat are a decisive factor when we choose our partners [. . . so] it seems fair to say that we possess a secret language of scent, and trees have demonstrated that they do as well'.[4]

From this perspective, there are many examples of active communication by trees, whether with each other or with other creatures. For instance, when acacias eaten by giraffes pump toxic substances into their leaves to deter browsing, they also emit a warning (in the form of ethylene gas) to signal to neighbouring trees that a crisis is at hand, and the forewarned trees do the same.[5] Elms and pines under attack from leaf-eating caterpillars call on parasitic wasps to prey on the infestation, again by emitting pheromones. Trees warn each other using chemical and electrical signals sent through mycorrhizal fungal networks at their roots. They use bright blossoms to attract pollinating insects. So 'trees communicate by means of olfactory, visual and electrical signals'. The existence of such communication amongst trees—including through fungal networks that have been described as a 'wood-wide web', is a frontier, if debated, area in ecology.[6]

While expanding how communication amongst plants takes place, these biological and ecological accounts offer little theorisation of meaning, however. Nor do they address human–plant communication. Indeed, while acknowledging, for instance, that humans can and do pick up on certain scents from plants—the fragrance of a flower, the smell of rotting bark—this is rarely construed as human–plant communication, because (as Wohlleben argues) the message is not aimed at or interpreted by humans in the way the tree controls or 'intends'.[7] However, as philosopher Michael Marder argues in relation to plants, intentionality can be understood as more relational and emergent, not as directed to a singular goal or to the conscious interests of a self.[8] There is no need to impute to a plant an intention to send people a particular message. A structural-biosemiotic approach goes further, as it downplays these questions of tree agency and intentionality and how much they matter, and reframes sensory crossover, pointing to structural ways that signs associated with senses of many kinds are picked up on by people within fields of signs, grammars, and through metasigns that convey packaged histories. To probe communication in this manner between individual plants and humans, we can usefully start in a familiar context of plant intimacy for many people—the houseplant.

Houseplant Conversations

Ethnographies beyond the human from across the pluriverse explore how people nurture plants in their houses, and in their conjoining courtyards and gardens. This is one focus in a growing body of human–plant ethnographies, addressing the challenge of drawing 'plants from the margin of [anthropological] research, without anthropomorphising them.'[9] These ethnographies enable us to see how plants and humans are entwined in relationships within which the responses of plants are agentive, without any need to posit a plant mind.[10]

For example, Catherine Phillips and Eily Schulz explore the nurturing of indoor plants in Melbourne, Australia, which is increasingly popular amongst young people.[11] Relations with cohabiting plants are shaped through everyday struggles to keep a plant flourishing, with the right amount of light, water and so on, in which living plants oblige the keeper in a reciprocal relation of care. People maintain these relations with plants as individuals (not just as examples of their species). Phillips and Schulz found that after keeping indoor plants young people in Melbourne 'could no longer imagine their homes without plants [...]. Caring for plants changes how homes look, feel, become, and are maintained as meaningful places.'[12] The emplacement of a plant has a reciprocal quality that transforms the dwelling, with houseplants blurring any boundary between indoor decorations and living beings. Extending their analysis, it seems that a flourishing houseplant becomes synonymous with wider flourishing and care, with human–houseplant relations becoming part of what Lestel would term a trans-specific, hybrid culture; in this case a culture centred on and helping to constitute a 'home'.

Human–houseplant relations can be intimate, as people consider their plants as friends or family. One young woman in Melbourne spoke of how indoor plants were her best friends and had 'saved my life'. Caring for them at a difficult time, when she was 'not looking after herself', led her to reconsider how she was treating herself.[13] Gideon Lasco found that during the Covid pandemic in the Philippines, nurturing houseplants became popular as a form of companionship. People spoke of their houseplants as kin, expressing familial bonds and considering themselves as plant parents.[14] In such examples of companionship and affection, human–plant relationships are not unlike those many people establish with animal companions. Likewise in urban Mozambique, Julie Soleil Archambault found that young men describe the ornamental plants they nurture in their backyards as their 'lovers'. She cites the example of Kenneth, who

> used to say that he could not afford a lover. In a sense, 'my plants are my lovers' was a joke. Yet at the same time, as I came to appreciate the affective bond between Kenneth and his plants, I also understood that when he described his plants as lovers, he meant that his plants commanded the same sort of time, attention, and affection that lovers normally would. The plants, in turn, loved him back through beauty and growth. In other words, his plants were quite literally his lovers.[15]

Archambault notes how human–plant relations take on new and more significant affective qualities in a post-socialist, post-war context in Mozambique marked by rising inequality and state retrenchment, as many of her informants understand and experience these intimate relations with the plants they nurture as more genuine and 'authentic' than love or sexual relations between humans, that are now so often almost commoditised. Relations with plants were somehow more genuine. Young men were highly reluctant to sell their ornamental plants; love, in this case, turned plants into inalienable possessions.

Others construe such plant kin as allies or as children. Michael Vine describes, for example, how in southern California, in spiritual healing practices, people are encouraged to associate with a particular plant 'ally' that accompanies them on their healing journey.[16] Theresa Miller found how Indigenous people of the Brazilian savannas consider the crops they tend by riverbanks and in forest gardens as 'children'—nurtured in caring, loving relationships. She discusses how kinship develops through everyday multisensory intimacies, as well as through mythical storytelling and encounters mediated through shamans.[17]

Whilst such works establish the significance of human–plant relations as forms of companionship, they barely probe their communicative dimensions. Some note the significance of multisensory signs in establishing and maintaining relations of love and care: the look of a plant that makes it 'stand out' as one to love, as a young Mozambican man put it; or, in Melbourne, the smells involved in an indoor plant's contribution to the making of a home, or the touch-signs of dry leaves or pot soil that signal its need for water. Picking up on such signs is helped by the familiarity and attentiveness that builds through co-living over sustained periods, whether in houses or indeed in other contexts such as gardening, herbal collection and more.[18]

A structural-biosemiotic framework can enable us to go still further than these approaches, however: to understand communicative dimensions and the meanings shared between people and houseplants as rooted in a structural semiotic encounter, not just experience of prelinguistic signs. To probe this, we authors can turn to our own houseplant experiences. We nurtured an indoor 'parlour palm' that had lived in a pot in our sitting-room for several years, until it started to show white powdery patches on its leaves and its lower branches started to wither, to dry and to brown. We read such signs (and metasigns) in the physical form, colour and texture of the plant, picked up through our human senses of sight and touch, as showing that all was not well, and responded—wiping the leaves clean, giving more water, moving the plant into a lighter place, and even interpreting the whiteness as fungus, spraying the leaves with chemical fungicide in some panic. It was not just the signs themselves we were picking up, but their conjunction, thus involving a syntactical element: the juxtaposition of signs was conveying a package of meanings of a plant 'not thriving' that could be read in terms of a structural grammar, as something out of place, out of the ordinary. There was thus a grammatical, even narrative quality to our reading of the plant—a story of transition from a subliminal orderliness and wellness to a problematic state. Signs of colour, texture, poise of leaves and so on made sense in relation to other colours, textures and leaf disposition linked grammatically and in narrative form. Our interpretation, in short, had a structural semiotic aspect to it, and this was the way in which interspecies communication between the plant and ourselves unfolded. It elicited a response—we did things such as watering the plant or positioning it differently; the plant's form changed; it responded to us and our care by exhibiting a different array of signs, in an exchange of meaning that was thus essentially a conversation. Whether or not the plant's part in the conversation was shaped by memory of past encounter is not clear to us, but we are interested to find that botanists now discern that plant responses can indeed be inflected by memory. Certainly we were learning through the responses of the plant, and in this way the conversation was little different from the kinds of conversation with horses and chickens that we have discussed above, inasmuch as plant responses contribute to something we perceive as an emergent code.

There is also scope for refining human–plant communication so that signs and their combinations are read more accurately. Sometimes expert knowledge of what a particular sign might signify for a particular plant species can be of help. Usually such knowledge is built through long experience. Some people just seem to have, or at least to have acquired, a particular communicative capacity with plants—captured in the English phrase 'to have green fingers', referring to someone for whom plants in their care just seem to flourish. Our own bungling efforts to save our parlour palm eventually failed: we never got to the bottom of what the white powderiness was, although recognising that a plant expert might have been able to do so. By contrast, when an enormous yucca tree in our kitchen began to show signs of unhappiness, its lower leaves successively drying out up the trunk, and seeming to do so more as we watered to compensate, interpreting this as a sign of dryness, we turned to the internet—driven by the size and grandeur of the plant (as well as the guilt associated with our palm failure). We discovered the symptoms to be a common sign in yucca trees of over-watering (not under-watering as we had assumed) and the resultant rotting at the roots. So we changed course, and now hardly water the plant; thankfully it is still thriving, and as we help it, it does not just decorate our kitchen, but contributes to the hybrid culture that is our home. This does not depend on anything prelinguistic—on signs signifying something simply through iconic or indexical relations as direct mechanistic reflections of what is happening to the tree. Something more structural is at work.

From this perspective, we need not contemplate whether or how the plant is 'thinking', at least not in a way that would impute to it an interiority or a mind. It is hardly different in this respect from much of the communication we pick up on from our human others in gestures and subtle expressions that we consider not as 'conscious thought' but, inferring from signs, as unintentional, subliminal, subconscious. This does not preclude us, however, from reading signs and inferring from them in ways that do invoke the feelings and emotions implied in some of the phrases used to describe plant (ill) health (and that might thus be seen as somewhat anthropomorphic). The floppy stem or drying leaf with its crispy ends that we pick up on through sight and touch might lead us to interpret the plant as 'thirsty', or 'unhappy'. When the thirsty plant is watered and perks up again, its form changes, and we interpret it as 'happy'. Observing a plant to be 'pot-bound' can evoke a feeling of boundedness and restriction in those who care for it, as it did for us; the metaphor invites mimetic experiential identification. When a nurtured plant dies nonetheless, humans can feel sad, or guilty, or incompetent, or more. All this involves a switch to a more anthropomorphic language, but one that captures phenomena that have emerged from interspecies communicative experience, virtually impossible to describe in words without appearing to anthropomorphise. Moreover, for human emotional connection, we do not know, or need to know, if or what the plant feels. This matters relatively little; it is the combinations of signs and grammar that generate the affection and connectivity that are, in turn, important to the function of care. It is the act of communication itself, not the imputing of mind, that matters.

The meanings communicated by houseplants extend well beyond those associated with immediate thriving, and over longer timespans; something we can again understand in a structural mode. A friend has a Swiss cheese plant, acquired during her student days and now enormous, and several decades and house moves later its continuing presence evokes for her that time, the person who gave it to her as a gift, and the continued social relationships that have also lasted; but such human social meaning is inseparable too from the decades of plant care involved. The plant effectively narrates the long, entangled history of our friend's life, through the size it has reached, the directions of its branching in moving from room to room, the scars where it has been scorched by a light.

Our friend's care for the plant in the present is shaped by this 'biographical' package of meanings, and so the plant has been responding, such that its current shape and dispositions are the result of this mutual history. In this way, a plant becomes a metasign conveying the history of its interspecies entanglements to those with the experience to appreciate it. Catherine Degnen captures this in suggesting that amongst gardeners in northern England, plants were 'autobiographical': 'particular plants became associated with specific events in a person's life such as a move or the beginning of a new friendship [...]. Each plant had its own story, even if some were remembered more vividly or fondly than others.'[19] What we would add is that the meaningful human social relations at stake are intertwined with the relations with the plant itself. Understood structurally, the disposition of a particular plant is a metasign that conveys a particular package of meanings in contradistinction to other plants that have a different significance (or none at all).

By tracing such communicative encounters in terms of structural biosemiosis, we thus become more deeply aware of the exchange of meanings that are part and parcel of practices of care, and the affection—and sometimes therapeutic qualities—that such exchanges involve.

Tree Communicative Companionships and Historical Entanglements

Turning from house and garden plants to trees, we can explore further how meanings are entangled with particular people and social events. Given the longevity of many trees, the package of meanings carried by a tree can tell of even longer histories: of successive humans who have lived with it.

For instance, as we walk through a patch of forest near the village where we lived in Kissidougou, Guinea, Lamin points out a tall cotton tree, extending high above the canopy. The surrounding forest is quite low and appears scrubby, its taller trees—of valuable timber species—having mostly been sold to loggers from town in recent decades. Botanists often read such tall cotton trees as relics of the gradually disappearing forest, tall because they originally grew up amidst other trees, reaching up for light. To villagers such as Lamin, and to us, anthropologists who have learned their perspectives, a lone, high cotton tree carries quite different meanings. Its emplacement next to a now abandoned settlement site, its tall form and lack of branches, are signs that this was actively encouraged as a lookout tree by the village's founders during late nineteenth-century warfare, taking root in the village's war fence. People encouraged it to grow tall by trimming its lower branches so that scouts could climb and warn of attack. It is distinct from the low-branching cotton trees, some of which still exist, that villagers produce by cutting the main stem, forcing them to branch into a shape that they call 'four arms' (*bolonani*) that intermeshes with other trees and provides a frame into which spiky lianas can be interwoven to form a war fence that would defend especially well against attacking cavalry. As we pass, Lamin recalls these stories of village foundation and protection, passed on through generations and significant to the political status of the village's current leading families. He reflects on the events that led the village to move to the roadside in the 1960s, leaving this old forest 'island' as a place of memories, ancestral graves and rich soils for coffee-planting.

Reading the archival significance of a particular cotton tree depends on an understanding of the conversational language (codes) established between people and cotton trees. The way trees accrete meanings related to the families who nurtured them, and become sites for offerings to important family ancestors or to the spirits they successfully negotiated with, depends on interspecies dialogue that can be understood structurally. Thus the form of the cotton tree in a Kissidougou forest island, picked up on visually by people, is a metasign conveying a package of meanings, in a field of contrasting metasigns: the tall form versus the short branching form; the cotton tree versus any other kind of tree. The trees convey the histories of their past nurturers and become a vehicle for the storying of family and political relations, encompassing settlement, alliances and rivalries. Every significant tree—every metasign—carries something of all the others that it is not: all the other possible other forms it could take, and by implication, of narratives that could be told. Metasigns acquire significance on the borderline between experience and narrative. Past conversations between people (within their social world) and particular trees shape the ways both tree and people develop, setting the stage for further conversations.

The contrast between a tall-form and a *bolonani* cotton tree is echoed in the contrast in English orchards between tall-growing fruit trees and those that have been deliberately pruned to take a short form for easy picking, or even trained to grow flat against a wall, in 'espaliered' form. The pear trees in our garden growing next to an old brick shed extend out from it now with branches in all directions, in shapes that convey an earlier history of being neatly espaliered by the diligent and tidy gardeners who lived here before us, and our own more lacksadasical approach and time-constrained life circumstances—as well as our aesthetic and emotional preference for these old fruit trees to grow freely, not rigidly tied down. The trees have burst out of their bonds.

In the public garden in our town, a large central tulip tree provided shade and atmosphere for countless family gatherings and picnics, and marked a site where parents and children would gather after school for conversation and round-the-tree play. When it eventually had to be felled because it had become old and dangerously weak, the sadness felt by many reflected the tree's capacity to evoke memories of such social times, and a melancholic sense of the loss of these—of time moving inexorably on. This is not just a case of a tree being used by people for human social purposes, for something slightly different and more reciprocal is at work. In a tree's growth and how it grows, and how we respond to it and it responds to us, as well as in how we live with it (play around it, sit beneath its shade) there is more of a conversation. If the tree goes well then life goes well (and vice versa). This is part of its package of meanings, underlining the poignancy of its final felling.

We have thus far been examining conversations with single plants, and how they convey meanings. These are not general to their species or variety (albeit general characteristics such as how a certain kind of plant reacts to particular soil or light conditions, styles of care and so on can be important). And in turn, knowing the individual tree and its particular life circumstances helps us interpret its signs. Yet as we establish more-than-human relations and intimacies with plants and trees, the communicative codes we develop with them are partly shaped by other relations, our own human social relations among them. In this, there is an interplay too with our particular framings of plant ontology: with what sort of a thing we consider a plant or tree to be, and therefore how we might relate to it. These framings give rise to different understandings of communicative possibilities; of which modes are important, and the meanings they convey.

In New Zealand, kauri trees, a species of conifer that are amongst the world's longest-living trees, are highly important to the identity of Māori people, many of whom regard them, given their age and size, as ancestors, linked with them through kinship. Particular tree individuals carry their own status and names; for instance, the largest known living kauri, which has been growing in Waipoua Forest in the north of the country for about two thousand years, is named Tāne Mahuta, after the god of forests in Māori thought. It carries specific meanings in Māori creation narratives, in which the god is said to have pushed apart the sky father and the earth mother to create space for life to thrive, and the tree is particularly special to some for this reason. Many kauri trees have been afflicted with a disease in recent years, in which a mould, phytophthora, attacks the roots, manifest in signs of lesions breaking out, and the tree exuding a yellow gum. Māori who identify closely with the trees as kin interpret these as signs of the tree attempting to protect itself with a thick armour, and thereby also as protection for themselves and their way of life. As a Māori spokesperson in Waipoua forest put it, 'The threat of kauri dieback to the species is a threat to Māori identity itself.'[20] It is this identification and connection that has underpinned strong collaborations between Māori campaigners and scientists in attempts to address the disease.

Whilst we can consider agency in such communication without positing a mind or 'modelling system', people nevertheless sometimes discern one—referring to a tree as if it were a mindful agent, as those in Australia do in saying that trees 'tell yarns', or to the spirit of a tree, or a plant, as Rastafari healers do when they attribute the agency, the efficacy, of a medicinal plant to the spirit (*gees*) associated with it. This might be considered as a particular framing of tree or plant ontology. It might also be considered as a way that people capture the communicative experience that is felt in their plant encounters and that provides a shorthand (sometimes elaborated) through which such experience can be related and acknowledged.

Emplaced Plants and Trees

In these examples, trees convey meaning not just through numerous signs linked to their form, but also through their emplacement. Where a tree is growing, and all its juxtapositions, is part of the syntagmatic dimension to the meanings it creates. An oil palm tree growing alone in a rice field in Guinea where farmers have nurtured it carries meanings different from those of one in the rows of an oil palm plantation; an oak in the centre of a public garden conveys meanings different from those of one in a forest. Trees can also be matter out of place: a large tree growing indoors in the centre of a public building or office communicates something that relates not just to its beauty or the feelings it evokes in visitors purely in itself, but also to their sense of surprise at finding it there, as if it were a work of art, calling upon us to interpret it in that way.

Indeed, whatever the framing of what a tree or plant is—its ontology—the meanings it conveys depend on its emplacement. As we saw for ornamental plants in Mozambique, it is their emplacement in a backyard that enables them to be 'lovers'. Intimacy is co-created with their emplacement; it is not just a matter of which species they might belong to. Likewise, medicinal plant specialists discern very different medicinal qualities according to the place where a plant grows. For example, Rastafari healers in South Africa seek specific plants not from cultivated gardens, where they are entangled with human affairs, but from the mountains where they are free from what the healers envisage as such compromise.[21] In Kissidougou, Kuranko healers identify different medicinal qualities in plants of the same species growing in the open sun, or in shade, or in the soils of a termite mound.

The emplacement of trees also has an enduring quality, linked to their stasis and longevity. This makes human relationships with trees significantly different from relationships with companion animals, all of which are mobile. Whereas we and our companion animals are all migrants, through conversations with trees we acquire our own emplacement and connection to place. Whilst we have noted how many animals 'write' on a place to mark their territories, our relations with trees mark territory too. Indeed they embody and become territory.

Foundational trees, established with a settlement by its founders, epitomise this sense of connectedness of people to place through trees; whether it be the cotton tree in central Freetown, in Sierra Leone, that was planted by the founders of this settlement for freed slaves and was mourned deeply by the city's inhabitants when destroyed by a storm in May 2023, or the now shrivelled ancient tree of Niani in the courtyard of a village in Guinea's dry savannas, said to have been planted by the founder of the Mali Empire. Such connectedness is found, too, where trees are planted or protected to signify settlement of a territory or mark its boundaries; a sense that extends from the mango tree planted in the centre of a newly built compound in Guinea to the way a newly moving-in British houseowner might plant an apple tree in their garden, signifying 'home'. Such connections have often been framed in legalistic terms: for instance, where trees are said to be owned by their planters and tree tenure is thus linked to land tenure, conveying ownership of the land the trees grow on, with immigrants or temporary residents often forbidden to plant trees, as this is said to convey an unwarranted claim to land. Yet such legalistic phrasing occludes how emotive affinity with a place also emerges through the emplaced conversation that nurturing a tree involves.

That plant and tree conversations emerge through involvement with a place in this way suggests that envisaging communication with a tree or plant independent of place may be problematic. To claim that one has or can establish communication with all trees, or all trees of a certain species (all oaks, or all cotton trees) is to appeal to a neoliberal supposition that plants are exchangeable, and transferable. This might be compatible with capitalist ontologies, but it is incompatible with emplaced, more-than-human experience.

Thus across the examples recounted above, we see how people intimately involved with plants in more-than-human worlds across the pluriverse—whether those with indoor plants, gardeners, those attentive to medicinal plants, farmers, those nurturing significant trees linked to birth, death, settlement history and identity, or indeed researchers—are affected by them and come to care about them. The social meanings that certain trees and plants convey, and the intimacy and affection that they can create, is also surely what underlies the passion with which campaigners protect particular trees from felling, whether for logging or road development: from the redwoods of California to neighbourhood oaks in the UK.[22] Similarly, this underlies the sense of loss and grief that can be felt if a significant tree is felled, and the passion given to retain and protect such trees, for tree removal can de-racinate its connected human inhabitants.

Forestry plantations, where trees are planted methodically in rows irrespective of surroundings beyond basic light and nutrient requirements, also de-individualise them, in estranging them from place and its associated meanings. They transmit a false impression that trees are more commodity than life-form, valued only for a single, narrow human purpose in market exchange, perhaps for timber, palm oil or carbon offset credits. This in turn permits an illusion of transferability: that trees felled here can be compensated by trees planted elsewhere—sometimes with the proviso that they will be of the same species or mix, as in the biodiversity offsets promised around mixed English woodland patches. We have here the tree equivalent of battery chickens. Yet just as each chicken is a life, so the place-rootedness and accreted meanings of trees deny the possibility of such transferability and replaceability. Those campaigning against new urban expansion in their areas are told that the groups of old oaks in the rural landscape will be compensated by new planting elsewhere in a biodiversity/tree offset scheme. Planners assert that a new town can bring 'biodiversity net gain'. Yet what we have tried to capture here is that this erases people's conversational entanglements with the particular trees they have communicated with all their lives; with landscapes they might not own, legalistically, but with which they have acquired an affinity. In a similar vein, when people in the village in Kissidougou where we once lived moved site to live nearer the road, they preserved the cotton and kola trees of their old forest island, with all its entangled meanings. The new village site would, in time, grow a new forest island intertwined with inhabitants' lives, but a different one, associated with place and connectedness for the families now living there, not a replacement for one lost.

6

Bees

Interspecies companionship and communication between humans and long-domesticated animals such as chickens and horses are relatively obvious relations to consider, and we have made the case too that such consideration can be extended to particular plants and trees. Now we probe ways in which people develop communicative companionships with insects, and what a structural-biosemiotic framework might add to understanding them. Our focus is on bees, and particularly honey bees.

We begin by reviewing emergent findings in biological research concerning communication amongst bees, and most particularly debates concerning their 'language'. This reveals the symbolic and syntactical dimensions to bees' communication with each other, and the ways the emergent codes are shaped both by insect memory and place. There has been debate over how far bee communication is innate (biologically programmed) or contingent on experience and social learning within the social world of bees, and thus whether one can speak of a hive 'culture'. Biological research has long treated bees as somewhat exceptional in their capacity for symbolic communication,[1] and such an approach has been mirrored in biosemiotics as well. Biosemiotics researchers, as far back as the founder Sebeok, have taken a more structural-biosemiotic approach to bee language, treating it as anomalously 'akin to' human language, in an way that they eschew with regard to other creatures. We are interested in the contradictions, inconsistency and instability that this reveals in the Peircean paradigm that dominates biosemiotics, and reflect on bees less as exceptional, than as paving the way for understanding other insects and wider naturekind.

Our main focus, however, to which we then turn, is on human–bee communication. Since bee 'waggle dances' have come to be appreciated, some bee-handlers have attempted to hack into them, and in doing so endeavour to frame their communication with bees in terms of bees' own language. Such cases are, however, recent and exceptional. Others have long been conversing with bees as part of collaborations involving mutual accommodation, benefiting from honey and the other affordances of cohabitation.

Examples we shall dip into show people picking up on a diversity of signs across the senses, whether scent, vision, sound or movement, that acquire meaning in relation to other signs and to their ordering or juxtapositions—their grammars. Equally bees themselves pick up signs from human scent, movement and rhythm. So whilst much of the communication amongst the bees themselves has been (and remains) beyond our perceptual reach, and whilst much of the communication amongst human handlers is outside the bees' sensory reach, there is enough sensory crossover and ability to pick up on packaged metasigns to allow for the codevelopment of shared codes and for ongoing conversation, however imperfect and contingent upon place and context. For people, such conversations have associated emotional dimensions involving identity, fear, pleasure and so on that cannot be divorced from relations of care, respect and more. While hesitant to impute human emotions to bees, those conversing with them are often highly attentive to what they read through metasigns to be the insects' collective emotional state—of anger, upset, calm, perhaps—that they seek to address.

Yet what becomes particularly interesting is the ambiguity about what exactly our companionship is with: what is it that are we communicating with—individuals or collectives? Contemporary science treats bees as 'social insects', with individual bees living as members of colonies with social hierarchies and relationships between their queens, drones and workers. Entomologist Thomas Seeley describes a colony as a 'harmonious society, wherein tens of thousands of worker bees, through enlightened self-interest, cooperate to serve a [...] common good'.[2] In this respect bee society becomes amenable both to social analysis and, potentially, to social metaphor. A bee colony or hive can be considered as a 'super-organism': as more than the sum of its parts, arranged into 'specialized castes that act together as a functional whole', as well as being 'self-organized by division of labour and united by a closed system of communication'.[3] Bees have been used as a metaphor to reflect not only on human social orders extending from communism to fascism, but more interestingly on whether social orders are limited to humans. Elsewhere in the pluriverse however, as we probe from the Philippines to Angola, we find radically different representations of the collective of bees and the place of people in relation to them.

Communication amongst bees is ecologically significant, as they are among the many pollinators that are so vital in their relations with the plants that sustain their colonies and the wider ecological order. The decline of bee populations worldwide, at times amounting to collapse, drives increased scientific and popular interest in bees, beekeeping and reflections on the entangled relations between people and bees. Yet as bees decline, we experience too an interruption to emplaced communicative orderings with wider ecological assemblages of which bees are an integral part; and this provides a bridge to our subsequent chapters that focus on communication with assemblages.

Bee Communication as Language

Honeybees use symbolic communication. Most notably they convey information about distant food sources or about possible hive relocation sites to other hive members through their 'waggle dance': as they weave their way across the comb and return, other bees can deduce the distance to recommended flowers from the number of waggles, and deduce the direction from the orientation of the dance and the return. Biologists interpret this honeybee dance, and, more recently, associated head-butting, scent release and wing vibration, as a multimodal language in which these signs combine to convey information. This form of communication amongst bees enables decision-making that is collective and strategic. For instance, when seeking to relocate, a hive will vote on the most viable site by the number of bees joining in the dance and the intensity of the dancing itself, and bees relocate only when a unanimous decision has been reached.[4]

Several aspects of the dance are taken to qualify it as language akin to that of humans.[5] First, there is a syntax to the dance language, with a regular set of rules enabling bees to convey with precision the distance, direction and desirability of a foraging site. Rules discerned to date include that the dance be focused on the resource most urgently required by the colony (whether nectar or water), that the dance inform about flower patches where these can be found, and that it should focus on the closest source that is rich and reliable. The waggle dance follows a stable code by applying such rules to the contingent needs of the bee colony. When honeybees use and follow the grammatical rules reliably when communicating, the result is very similar to human language. Scholar of science, technology and society Eileen Crist elaborates on this parallel. First, just as observance of rules in human interaction implicates communicative competence—a fairly effortless capacity to follow shared rules, without any reflective knowledge of them, being all that is required—so it is with bees. 'The extrapolation of a rule-set for the dance language does not imply that the honeybees are deliberately following rules, only that they can be seen to abide by them and use them competently.'[6]

Second, bees deploy symbolic communication, in which one sign or set of signs can stand for something else: a truly symbolic message 'which is separated in time and space from both the actions on which it is based and the behaviours it will guide'.[7]

Third, bees communicate pragmatically, adapting their dance to social need and environmental circumstances. The dancers re-choreograph according to their experiences of changes in quality and quantity of available water and nectar and yet are sensitive also to the feedback they receive from fellow workers, modifying communication accordingly. A dancer who senses (somehow) that the resource they know is not as desirable or necessary as another's will stop dancing for it; dancers 'listen to the "applause" of the unloaders'.[8] Bees thus adapt their application of linguistic rules to changing situations, enabling collective decisions.[9] A fourth linguistic dimension is that the dance does not have just referential qualities, but performative ones: it calls on other bees to act, to do something—to 'fly there for food', or to 'relocate here'. By suggesting action that follows its message, the waggle dance is, again, akin to human language.

As the biological revolution unfolds, the linguistic shaping of bee society is being understood as far more complex, in its multimodal quality and the array of meanings conveyed, than was previously thought. Dances extend from the waggle we have described to the shake dance, tremble dance, whir dance, grooming dance, massage dance, alarm dance, joy dance and so on, the scope and meaning of which have only begun to be deciphered.[10] Questions have arisen concerning the extent to which such dances and their specific symbolism are innate or are socially learned and context specific. Most recent research finds that whilst aspects of this communication may be biologically programmed, it is always modulated by social learning, and consequently the innate dimension becomes an anchor in fields of meaning that are shaped socially and by experience. So whilst bee communication per se has usually been assumed to be innate to the species, as 'the language of bees', the dance is honed for accuracy as young bees observe experienced bees. Untutored bees cannot understand distance coding at all.[11] The existence of such social learning thus allows the development of hive- and place-specific traditions or cultures in much the same way that human linguistic capacities allow for society-specific languages and dialects.[12] As science writer Philip Ball puts it, 'The waggle dance code has regional dialects that might take into account the local terrain.'[13]

That bees meet criteria through which honeybee dances qualify as language is widely accepted by scientists, and is celebrated as a source of wonder by some; but not all are without qualms. Many find it hard to allow such 'human' capabilities to be attributed to an insect[14]—situated so far down the evolutionary hierarchy of beings given the lower-order status of insects in established scientific classifications. In biological research bees have come to be regarded as somewhat exceptional in their capacity for symbolic communication. Martin Lindauer noted that 'there is no form of communication in the animal kingdom comparable to the dance of the bees'.[15] Yet such bee exceptionalism is partly an artefact of the long-standing scientific attention focused on these creatures. The technological and biological revolution has now enabled communicative practices to be found in many other insect and wider arthropod species, and this is a development set to continue. In the many social insect species investigated to date, although innate responses form a basis for communication, social learning is always a critical modulator of communication processes.[16] Emergent codes are always shaped by memory and emplacement.[17]

The linguistic capacity of bees has also been treated as somewhat exceptional among those probing biosemiotics. Biosemiotics researchers who investigate bees, going back as far as Sebeok, have long considered bee language to be like human language to an anomalous extent, and have been induced to take a more structural-biosemiotic approach to it than they have applied to other creatures. Focusing on the waggle dance, Sebeok recognised honeybees in some of his earliest work as 'the semiotic species par excellence, possessed, next to our own, of the most elaborate social communication system thus far recognized by ethologists'.[18] In his subsequent works, however, he, and the many who followed him and whose work we discussed as we developed our analytical approach in chapter 2, directed attention away from symbolic communication to focus on what they identified as prelinguistic iconic and indexical modes of communication, eventually excluding the possibility of symbolic communication and code in non-humans. We have found it hard to square Sebeok's early attentiveness to bee language with his later insistence that non-human signing was prelinguistic, and this reveals a degree of contradiction and inconsistency within the field of biosemiotics. In both biology and biosemiotics, however—and in line with what our entire book is probing—bees' symbolic communicative capacity, instead of being cast as exceptional, should rather perhaps be taken as a stimulus towards discovery and reflection that can be extended to other insects, and indeed, to wider naturekind.

Human–Bee Communication

Biological research examines how bees communicate with each other, and reflects on the methods whereby scientists have come to perceive this. Our focal interest is rather different: it is on human–bee communication. Popularisation of scientific understanding of the bees' waggle dance has led some to attempt to enter communication in the terms of bees' own language. This is a recent, and exceptional, approach, however; in most parts of the pluriverse people have long been conversing with bees in other ways.

Those beyond researchers who have tried to interpret and 'hack into' bee dancing have generated a plethora of web resources aimed at laypeople wishing to keep bees. One self-styled 'bee whisperer' in the United Stated has a website and a YouTube channel that hosts a basic video on 'how to speak bee', along with a range of more specialised films on aspects of bee management, whether building up a colony, preventing swarming or establishing a brood chamber.[19] The video provides detail on the relationship between shape of dance (round, sickle, hybrid), the relations between the waggle timing and vigour and the distance and qualities it evokes, allowing that these precise details vary according to the strain of bee, with 'the Italian' being the kind most commonly kept in the United States. Bees, like humans, it is suggested, have different cultural dialects. Beekeepers can observe these dances and so 'pick up what bees are saying', which is of practical use in knowing exactly where bees are feeding.

Another site, *PerfectBee*, also describes how bees 'talk to each other', and urges people to learn to understand this—not just for practical, but for emotional reasons:

> Beekeepers who understand the ways bees communicate with each other and with us will likely be more successful and, in my opinion, enjoy the beekeeping experience much more. For me, understanding how they talk to each other makes looking in the hive that much more exciting and fascinating. And understanding how they talk to me means I'm able to interact with them in a way that keeps them happy and me unharmed [...]. Honey bee communication is on par with much more advanced animals [...]. It's just one of the many reasons I love beekeeping and feel inspired each trip to the apiary. I hope you'll enjoy it as much as I do.[20]

Understanding bee communication is thus proposed as a route to greater enjoyment and to a caring relationship between bees and humans. Yet while human understandings of bee communication might enable humans to live better with bees, we cannot really know the details of the meanings their communicative grammars and performances convey. Through dancing bees actively negotiate and renegotiate their social life, but as Crist puts it,

> neither its informative content per se, nor its phenomenology, can reveal to a human perspective whether the force of the dance is an order, or an entreaty, or, for that matter, something that no human word exactly translates. If honeybees do speak, it is also the case that we do not fully understand them: the et cetera clauses of their dancing are, for the most part, an obscure affair.[21]

As we have been arguing, this is not so different from most of our communication amongst ourselves or our interspecies communication with other animals, which tends to be a matter of pragmatic accommodation and getting along, without being able (or needing) to penetrate the details of our interlocutor's thought.

Beyond the waggle dance, educators also draw attention to other sensory modes through which bees communicate. For instance, *Perfect Bee* discusses sound and stings:

> The bees do talk to us [...]. That buzz noise we all associate with bees? It changes depending on the bees' mood. If the bees are feeling threatened or distressed, it increases in intensity and volume. When the bees are calm, it is quieter and slower. The first thing I do during any hive inspection, or any other time I approach the hive, is listen. If the bees sound 'happy', I know I can proceed without concern. If they sound 'angry', I may elect to return another time. Disturbing already anxious or upset bees is a sure way to get stung. [...] And speaking of stings... That is, in fact, another way bees talk to us. Honeybees only sting when they feel threatened. If they perceive that either brood or food may be at risk, they will sting. They'll warn us first with the changing in buzzing just mentioned, but if we don't listen, they sting us. Each female bee stings only once and dies as a result. If we heed the warning and retreat, we won't be stung after that. Honeybees, unlike Africanized bees, don't chase or follow. Once they know the danger is moving away, the go back to work and gradually calm down.

What is described here could be understood as a kind of unfolding conversation, which proceeds in structural ways. The bees signal through a buzz that acquires significance to the bee guardian in a field of meaning: 'this kind of buzz', so not another kind. The bee guardian responds through bodily positioning, approaching or backing away, and quite probably some subliminally perceptible human movements too. Mis-approaching might lead to a sting, conveying meaning to the human guardian that they have overstepped a mark, and then eliciting a human response involving changing position: backing off.

Biosemiotics scholar Heidi Campana Piva draws attention to the importance of smell in human–bee communication.[22] Focusing on 'stingless' bees (tribe Meliponini, which do not waggle dance), she documents how they sign to each other in olfactory/chemical modes, by leaving trails of odour spots to mark the way to feeding sources, and that summon others to particularly nutritious plants. Smell allows for shared communicative codes to develop between bees and humans, since this is a mode in which there is enough sensory crossover.[23] In Brazil, Indigenous Kayapó people also 'follow the odour that bees used to mark nesting sites'.[24] Barbieri, a bee-handler and researcher in metropolitan Brazil interviewed by Piva explained how the bees he cares for distinguished him from others by smell, and were calm around him: 'As they get used to your presence and smell, they stop identifying you as a threat.' Such a capacity to distinguish between individual humans has been shown for other insects, such as cockroaches. Piva suggests that 'if there is any possibility of a shared code between humans and bees, it can be through the use of olfactory signals'.[25] Whilst she draws on a Peircean biosemiotics perspective, and distinguishes these olfactory signs as 'more indexical' than the ones used by honeybees, which she denotes as 'more symbolic',[26] we question the relevance of such distinctions, as they obscure how olfactory signs might convey meaning in relation to others (and absences, and non-olfactory signs) within structured fields of meaning. Whilst we are not specialists in Meliponini, it would surely be premature to characterise scent as indexical without attending more explicitly to the contrasting meanings between one smell and another (or no smell), as well as to the ways scent might be integrated in multimodal signing.

Piva does, however, consider how meaningful communication between Barbieri and his bees has built up through mutual experience: codeveloping a conversational routine. This depended on an established sequence of weekly checking, opening the hive boxes, to which both handler and bees had become accustomed. As Piva writes,

> According to [Barbieri], when it is time for the handling routine, the older bees leave the hive box and wait outside, while the younger ones retreat deeper in the hive to hide from the light. As long as the routine is kept, and the bees have this constant contact with the human who is doing the handling, they do not present any sort of hostile behaviour, indicating that they are indeed used to the happenings.[27]

She notes that it is through past experiences with their handler that signs emanating from him (of movement, bodily placing and touch, as well as smell) have become meaningful to the bees, signifying 'not a threat'. This, in turn, has become a habit, in which the bees accept the intruder into their space without swarming or attacking. As Piva notes, the result is that these bees recognise their handler, regardless of whether this is conscious. In our more explicitly structural-biosemiotic approach, what is being described here we might consider as a conversation, conducted according to shared codes built up to integrate multiple signs into packages of meaning and functional grammars.

Just as communication amongst bees involves emplaced contexts and cultures, so it does for human–bee communication. The example of Barbieri and his bees shows how human–bee communicative modes build up. Anthropologists Lisa Jean Moore and Mary Kosut also illustrate the emergence of interspecies human–bee culture in the differently emplaced context of urban rooftops in the New York metropolitan area.[28] Here honeybees have long been cohabitants, but their presence has grown recently amidst revived interest in urban farming, local food, green consumerism, DIY culture and a demand for gourmet 'boutique' honey. As Moore and Kosut describe it, just as 'the city swarms with human activity, these bees buzz as they pollinate fruit, vegetables, plants, and wildflowers, playing an integral part in local urban ecology'.[29] The honeybees' communication with beekeepers is structured by its emplacement as 'these urbanites manage the hives in their backyards, community urban gardens, apartment rooftops, or attached decks, wherever space and access allow'. Urban beekeepers pick up on and respond to bees' signing, in conversations in which 'being with the bees involves smelling, hearing, tasting, and feeling them within the human body'. One urban beekeeper explains that 'there is something very Zen about the process of taking care of the bees. You have to slow down and really pay attention because you don't want to upset them and you don't want to get stung. When inspecting a hive, your mind can't be anywhere else, it has to be on the hive, observing exactly what is in front of you.' Urban beekeeping is 'a responsive performance of mind/body and bee', in which,

> much like a dance or practicing Tai Chi, there is a beauty in the movement and flows—choreographed and improvised at the same time [...]. The performative nature of beekeeping calls for embodied learning and sensitivity. Here, the bee becomes the educator/teacher through a co-mingling and penetration of the senses. Becoming attached and in sync with a colony or a hive is a ritualistic process, but it is also a sensual one where insects and humans connect, overlap and collide.[30]

We might add that such embodied learning is as embodied as language learning; indeed it is language learning for the humans involved. There is also an aesthetic and affective significance to this kind of urban interspecies communicative companionship.

What we see in these examples are shared codes, always in development and always emplaced. Much of the communication amongst the bees themselves is beyond our perceptual reach, just as communication amongst the human handlers is out of the bees' sensory reach, yet there is enough sensory crossover and codevelopment of shared codes, often operating through metasigns and their juxtapositions, to enable ongoing conversation, however imperfect.

For the humans these conversations have an affective quality—whether, at different moments, the emotion is fear, pleasure, an aesthetic sense, or something else. People's participation in these conversations is also attuned to, and responsive to, what they pick up on as the bees' emotional state: whether at a given moment they seem angry, upset, calm, or otherwise, and as they respond, this drives the conversation on in particular directions.

Conversations and companionship differ according to how bees are construed and framed. The conversations under consideration here are less with individual bees than with a collectivity of a colony or hive. Yet in saying this, we should be aware that we are still framing bees in ways compatible with scientific understandings of them as 'social insects': self-organising collectives with their own internal social hierarchies and 'hive mentality'. People remain separate from, yet might communicate with, the collectivity, albeit in ways that sometimes allow for a commingling of sensory experiences.

Elsewhere in the pluriverse, however, this is not always the case, and it is instructive to understand human–bee communication in settings where humans and bees have longer cohabited as part of shared social and ecological orderings. Here we find equivalent but differently framed representations of the collective of bees, in which humans are not so separate from—indeed, might be envisaged as part of—the collectivity.

Amongst inhabitants of the Indo-Malaysian Pälawan region, for example, human sociality and the honey-gathering that helps sustain it are inextricably entangled with bee sociality. Maintaining these entanglements as respectful and therefore productive is enabled through ongoing conversation amongst humans, bees and also the bees' own 'authority figures'. As anthropologist Dario Novellino writes, bees (like many animals) are imagined to have their own Master—a 'mystical' beekeeper in charge of their welfare called 'the owner of the flowers' (*Ämpuq ät burak*) or 'the Master of bees' (*Ämpuq ät mugdung*).[31] This Master, considered a kind of person, is said to dwell in an upper world, and can be seen only by shamans during trances. The gathering of honey from the hives and nests that various species of bee make, often in tree trunks, is a significant part of local livelihoods and human social life. It usually involves harvesting a hive and evicting its inhabitants. Yet this is not undertaken lightly, but rather through a set of negotiations or consultations with the Master of bees and with the bees themselves, aimed at maintaining a respectful relationship that is central to the social-ecological order in which humans, bees and plants are embedded. Consultations unfold partly through everyday practices in which hive gatherers climb vines to reach the tree canopy, where they use smoke as a sign to drive away the bees. Yet they also prepare new shelters to attract and encourage bees to construct hives (moving branches, logs, vines, mosses, spider webs and dead leaves) and informing the Master of flowers so that they can welcome the bees. The hive is then cut, wrapped in leaves, placed in a container and lowered down with a rope, destroying the nest. This dynamic of hive/nest destruction and construction, driving away and welcoming, unfolds in conversations between people, bee swarms or colonies, and the Master.

Whilst biological science considers bees in relation to a higher order 'hive', people in Pälawan frame such higher order relations in terms of the Master of the bees; a framing through which to help interpret and understand them, and with which to communicate. Yet in Pälawan this higher, coordinated social order is not simply of bees but involves humans too. Bees inhabit the upper levels of the universe, visiting lower layers to obtain food in seasonal cycles that account for the seasonality of honey production. People sustain this ordering—and its wider importance in terms of 'helping the earth', as many in Pälawan say—by performing a great offering ceremony every seven years or so. One shaman explained that this offering

> is made to ask the God creator [Ämpuq] for the flowering of trees and the growth of rice. Only if plants are flowering the *mugdung* bees will smell the fragrance and come down to the *tängäq tängäq* [the middle level of the universe]. We make an offering to the God Creator, and the *bäljan* [shaman] will ask the Master of flowers to forward our request to him. If the request is accepted, the harvest will be beautiful, the flowering of the trees will be beautiful, everything that God made will be alive. When this happens, the bees will give abundant honey.[32]

Gatherers wear charms (*pängtiq*) when honey gathering, whether those carved from the wood of bees' favoured trees and decorated to resemble the cells of the honeycomb that help the gatherer deal with swarming, or others of the *äjaq-äjaq* plant which weaken the bees and reduce their aggressiveness. They might leave a fragment of glass on the hive, 'as an eye, to see the honeycombs'. Integrating technical and symbolic dimensions in ways we cannot decipher, such charms can be understood in a structural-biosemiotic framework as signs that acquire meaning within structured fields: that become part of human–bee conversation.

Turning to Angola, João Afonso Baptista captures how human and bee society merged for a group of war survivors who found themselves squatting and hiding on others' lands, and who attribute their survival in such refuges to the bees that sustained them. Through these people's ingestion of their honey, bees embraced them to the land and earned their respect. 'The bees make our bodies,' as one informant said, for the bees draw all from the land, so by ingesting their honey 'we ingest lots of land, and we stay strong and safe'.[33] Bees, in short, conjoin people with land, and with place.

In parts of the pluriverse such as this, more-than-human relationships with bees, and the structured communications that underpin them, have long been a central and visible part of everyday life. For others, bees, their communications, and their ecological significance have been more 'under the radar'; taken for granted perhaps, less visible, only apprehended through a glance or a sting. For some, bees are coming to greater notice, as they are nurtured to become more numerous and central cohabitants of our cities, and as scientific and media commentary dwell on them. Ironically, it is in part as bees disappear, perhaps—threatening the ecological ordering that they are central to—that they become newly visible to us.[34] Bee absence becomes a kind of matter out of place that in turn drives increased human social attention: to the significance of bees, and to their potential to contribute meaningfully to social as well as ecological life. Whilst bee decline has been attributed to a wide variety of causes, attention to communication already suggests that environmental pollution has led to cognitive impairment.[35] We might add that in the more-than-human social world of which we are a part, such impairment derives also from conceptual pollution; from mechanomorphic ideas that annihilate communicative possibilities, and the recognition and care that might follow from these.

7
Bats

We now turn to relations with some of the world's 1,400 species of bat, our final case of communicative companionships.[1] As with bees, we consider human interactions less as with individual non-human animals, than as with collectives. In contrast to their relations with bees, however, people rarely attempt to develop language with bats and to establish conversations, and nor are companionship relations about livelihoods and extending capabilities, or not in quite the same way, at least. Instead, considering bats reveals ways in which languages—signs in fields of meaning, their grammars, and their packaging—can be significant to mutual accommodation; to enabling people and bats to live alongside each other in ways that are intertwined and mutually ordered at a higher, landscape level. This consideration provides a bridge to some of the examples we delve into more deeply in the chapters that follow, where we consider communication in and with wider assemblages in naturekind.

This chapter takes up three sets of issues. First, bats have become iconic in the biological revolution, given the discovery of their sophisticated sensorium linked to echolocations and the associated signalling and information exchanges. This entirely different sensorium has engendered classic debates probing 'what it is like to be a bat' and exemplifying concerns with *umwelt*. It allows us to consider debates about multisensory communication and the relationship between so-called prelinguistic and linguistic communication, and how both biological and biosemiotic conceptions have reinforced framings of bats and humans as radically different, leading lives as separate as their respective *umwelten*.

Second, we turn to parts of the pluriverse where human and bat lives are entangled, oblivious to concerns with echolocation. We show how a structural-biosemiotic analytic can capture broader manifestations of human–bat communicative encounters in worlds in which bats are part of human life and humans are part of bat life. Such entanglements are highly significant given that humans have co-evolved with bats, our ancestors having lived in closer proximity to many species of bats than most of us do today. Whilst we are ever more invasive of much bat habitat, accentuating fears of contracting novel zoonotic viruses from them, we should not overlook the times when, and places where, people shared more proximate relations with bats, including in the deep caves and rock shelters that were our ancestors' homes. After we built our own 'caves' in the form of huts, houses, temples or factories, many bats made the move to cohabit with us. In zoological parlance, many bats are therefore synanthropic species. Nevertheless, and in contradiction to this, long-standing mythologies concerning devils, witches and vampires, and emerging policies concerning conservation and the viral spillover of zoonotic diseases, are dominated by conceptualisations of bats and humans as separate and to be separated; overlooking the interactive lives that we show to be communicative.

Third, one consequence of human evolutionary cohabitation with bats, and the lives intertwined with them that some people still lead, is the exchange of viruses that we can consider as a dimension to this communication—as a feature of subliminal communicative orders of which we are unaware, but which influence our health. This brings us to what Peircean biosemiotics view as 'endobiosemiosis', and to reflect on its potential reframing within a structural-biosemiotic paradigm. In transcending both scales—from bats in landscape ecologies to bat viruses in bodies—a focus on language beyond the human reveals how attempts to separate and divide, to unravel human–bat entanglements—might be not just futile, but also harmful.

Communication Amongst Bats and its Sensory Sophistication

Bat communication has been reconceived in the biological revolution. The puzzle of how bats apparently navigated in complete darkness was probed from the early twentieth century and the scientific discovery of 'echo-location' or 'echo-perception' in 1944, provided a solution. Bats are able to fly, navigate around objects and even catch insects such as moths on the wing by emitting ultrasound and listening to the echoes; a mode of perception that enables them to pick up on the position and texture of objects, and the beating wings of insects and of other bats. Bats thus pick up signals from—communicate with—each other, with other creatures such as insects, and with the world around. For instance moths are able to pick up this mode of signalling from bats and indeed have developed a system of ultrasound clicks of their own. Emitting these in the last second before a bat attacks, they provide a 'blocking signal' that leads the bat to misjudge its distance, spoiling their hunt;[2] an example of non-human animals using communication to deceive.

This sensory world of bats that is closed to us has become a focal case in the philosophy of ontology,[3] and in biosemiotics, as an exemplar of the concept of *umwelt*:[4] the particular semiotic world of an organism that comprises its sensorium—what it can sense—as well as all the aspects of world around (food, shelter, potential threats, points of navigation) that are relevant for it. Viewed from a Peircean biosemiotics perspective, an organism creates and reshapes its own *umwelt* through its prelinguistic signing, that interacts with the world in a 'functional circle'. Bats' ultrasound might thus be conceived as a form of indexical signing that interacts with and might influence and reshape the state and locations of the objects, insects, and food sources that, along with its conditions and rhythms of dark and light, are part of the bat's *umwelt*. That a bat's *umwelt* is so particular, contrasting so strongly with that of other creatures, has become a case in point for the notion in biosemiotics that multiple *umwelten*—multiple realities—can coexist within the same environment.

Notably, bats' semiosis through their echo-perception is inaccessible to humans (as ultrasound is of higher pitch than humans can hear), so the bats' rich sensorium is to humans just quiet darkness. Nevertheless technologies now enable people to pick up on bat echo-perception through ultra-sonic audio recording of calls that can be heard (when remodulated) or viewed on a spectograph, combined with deep learning algorithms to locate their origins. Such technologies have been harnessed by researchers and 'citizen scientists' to track and monitor bat populations and their relationships with changing ecologies. Those developing the tools with volunteers across Europe find that humans are adept at distinguishing between a bat and a non-bat call, different types of calls and what sequence a call belongs in. The aim is for more people to become appreciative of the complexities of bat lives, their value and how they are being affected by global change.[5] As ways of studying communication amongst bats, it is notable that these technologies are picking up linguistic properties such as the importance of sequence and syntax, as well as differences in bats' acoustic signalling not just according to species, but also according to place amongst the same species. Bats of 'The Norfolk Broads' or near and far from roadsides, for instance, are characterised by different patterns of acoustic signalling—suggesting the importance of emplacement and context in their communicative codes, and thus the significance of social learning and dialects.

Thus the biological and technological revolutions are contributing to a conception of communication amongst bats more as language—possibly associated with particular, emplaced bat cultures. This contrasts with—and will increasingly challenge—the emphasis in biosemiotics on bat signing as prelinguistic, albeit in relation to particular *umwelten*. There is certainly a case for better integration of these contradictory theoretical perspectives on how bats communicate amongst themselves, and one to which a structural biosemiotics could contribute, highlighting the importance of grammars, codes and emplacement in the creation and sharing of meaning amongst bats and with their surroundings. But however conceptualised, the focus on echo-perception in communication amongst bats is a focus on what is largely inaccessible to humans (except those with specialised technologies) and arguably this has contributed to a conception of bats as ontologically different from humans, occupying different (and potentially) separable worlds. The problems with this position are revealed as we start to explore settings where human lives and bat lives are inextricably entangled.

The Structural Biosemiotics of Human–Bat Coexistence

Echo-perception is a relatively recent discovery (and not ubiquitous to bat species); people meanwhile have had communicative experience with bats through other sign systems that a structural-biosemiotic analytic can illuminate. In settings where humans and bats are cohabitants in everyday life, human lives and bat lives coexist and codevelop in ordered ways supported by more-than-human communication.

We can start downtown in the city of Accra, Ghana, where a colony of a million or more fruit bats inhabits the canopies of the cotton trees in the compound of the government hospital. By day they hang in the trees, conveying their presence through the black density they bring to the canopy, and the slightly pungent smell that permeates the air below; signs picked up, although rarely commented on, by the health workers, patients and visitors who go about their business underneath. The bats are just there, literally hanging out, with one or two occasionally flying down to feed on an orange or banana dropped by a passer-by; they are a part of the order of the place. As dusk falls, however, the density thins out as a part of the population takes to the air, circulating for a while and then starting to fly en masse in a single direction, towards distant feeding sources. Whilst ecologists have been tracking the specifics of their ranging and rhythms using radio-transmitters, human cohabitants of the hospital and city experience the movements in other more structurally communicative and multimodal ways.

They pick up on the visual contrasts, between the dense black and then fragmented tree canopy that bats have vacated; and between a clear sky, and then one speckled with flying bats: some of many permutations of canopy and sky appearance. There are sound contrasts too, as the relative silence of hanging bats gives way to a plethora of squeaks and flapping wings. The signs are read as packaged to convey a meaning of bats 'on the move' versus bats 'still'. They are ordered within a grammar that is emplaced both physically—in the heart of the city, with trees and bats contrasting with the streets and buildings around—and temporally, with the rhythm of the bats leaving in the evenings and returning in the mornings, or leaving in one season and returning in another. This is part of the ordered rhythm, often unspoken, of daily and annual cycles. Bats and their comings and goings are thus part of the multispecies ordering of the hospital compound. Occasionally those feeling inconvenienced sign to scare them away, and soldiers might let off a round or two of gunfire to this end, but to little effect. Whilst this can be construed simply as an interactional ordering of mutual accommodation, mutual cohabitation, it is also a communicative one. Moreover it is a structural one: if the bats did not take to the air or did not return it would be noticed as a disruption; as matter out of place. Indeed the presence of the bats has acquired a series of further meanings for some people involved, as it is associated with the powers of historic chiefs, lending added significance to their absence.[6]

If we travel north to rural villages in Ghana,[7] or indeed to other villages in West Africa's forest zone, including those where we authors have lived in Guinea, we find a similar intertwining of bat and human lives. Bats of various species inhabit the canopies of courtyard trees. They inhabit the roofspaces of houses and outdoor kitchens. They inhabit islands of forests that surround actual or former villages, and sometimes the caves where people take refuge in times of war. People are aware of their presence, picking up on visual and sound signs both when they are still and when they fly around at night, when the rustle of bat wings in house rafters or the silhouette of a bat against a moonlit sky is a feature of the diurnal rhythms of village life.

A villager in Ghana spoke thus of bats in the Tano sacred grove: 'In the evening around 4 or 5 p.m., you see the bats leaving the forest and flying over the town to go and feed. The sky becomes black with them. At dawn, they start to return around 4 a.m. When you hear them crying and making noise, it means they are returning. By 6 a.m. almost all of them have returned to the forest.'[8] Residents even use this rhythm to determine time. But for the most part these movements go unremarked; bats are just part of daily life, and of the order of things. Sometimes (such as—and perhaps only—when prompted by a researcher), villagers will comment on the ways bats share food with people, liking the same kinds of fruit and picking up their waste; and on the ways bats thus distribute the seeds of fruit trees, or indeed help with seed dispersal in village forest islands. Sometimes there are direct interactions: for instance, when people consume or trade certain species of bat as food or incorporate them into medical practices.[9] There is more generally an acceptance and appreciation of bats as being an inevitable part of more-than-human assemblages, and village and peri-village landscapes, both in the present and historically. This involves living alongside each other, accommodating each other; a mutual tolerance in cohabitation, without very much mutual exploitation or direct interchange. This is rather different from many other human–animal companionships, which are very much about direct interchange, and extending capabilities. There is nevertheless an interchange of affect: villagers like having bats around, as a reassuring presence that is part of the appreciated multispecies sociality that is village life, and part of a healthy, cyclical world. Times of day and season are signalled by their comings and goings. Bats are appreciated for their presence, and would be noticed for their absence—absence is matter out of place, or disorder, signalling an interruption to a spatially and temporarily syntactical world within which bats are central to the making of meaning. Many bat species, for their part, seek out human-inhabited landscapes in which to live; something that could be attributed narrowly to shared food sources, but plausibly, more broadly, to a comfortable and appreciated experience of and security in shared dwelling. Humans and bats can perform this accommodation without needing to understand much about each other—without imputing anything about interiority or cognition, and without the kind of communication that relies on sensory alignment. Rather communication is enabled by more-than-human language that is structural and emplaced, enabling the mutual respect for and coexistence of ordered human and bat lives. Different bats and their habits acquire particular social meanings for people: for instance, in debated narratives that link the presence of bats to the significance of shrines to ancestors where bats live, and associated claims of chiefly power;[10] but what is important here is that these human social meanings are entangled with exchanges of the more-than-human social meanings with bats themselves.

Across the pluriverse, many species of bat live amongst people's habitations and amidst the human-influenced ecological assemblages in which they can more easily find food, shelter, protection from predators, and perhaps fulfil a range of other needs and desires. Whilst in some settings mutual accommodation prevails—underpinned by respect for each other's orderlinesses and grammars—human–bat encounters are also framed in a range of other ways. Thus in European settings where insectivorous bats cohabit churches, barns and houses old and new, they often come to attention in the course of more bureaucratic encounters in which property developers and building restorers grapple with those legislating for bat protection. Other contemporary discursive framings and debates focus on the ecological and economic value of bats beyond communicative relations: ecological in terms of their maintenance of biodiversity, and economic in terms of resulting 'ecosystem services' and 'natural capital', as they might be construed. This highlights the role of fruit bats in pollination and seed dispersal, and of insectivorous bats in controlling insects, as, 'silent and often unnoticed, bats work the graveyard shift and help keep our world pest-free and biodiverse'.[11] This enables bats to be financially valued:

'Whether eating their body weights in insects every night, or dispersing seeds from fruit trees across large areas, bats provide services to local economies worth billions of dollars across the world.'[12] Despite the emplaced, multifaceted communicative entanglement of human and bat lives, such framings consider bats in terms of particular, divisible roles and values; framings compatible with the commoditisation and exchange of 'units' of bats and bat services, as one finds, for instance, in emergent biodiversity credit markets. Insofar as they overlook the affective communicative order, of course, such framings are also compatible with physical separation of bats from people.

Even where bat synanthropism, as opposed to separation from humans, is accepted, it is easily instrumentalised to human ends. For example, human–bat entanglement was once deployed militarily, when in 1943 US scientists developed 'bat bombs' that were supposed to create firestorms in Japanese cities. Each bomb packed in a thousand or more Mexican free-tailed bats that had been cooled into hibernation and each had a small, timed incendiary device attached. The bomb was to be dropped at dawn with a parachute opening in mid-descent, when the bats, by then warmed up, would be released to disperse. The idea was not that the bats would disperse into the Japanese wilderness, separately from human worlds, but that as night became day they would head directly to the eaves and attics within a twenty-mile radius. Safely at roost, their suicide belts, set on timers, would ignite in the inaccessible rafters of the flammable wood and paper buildings of unsuspecting Japanese town and city dwellers. Napalm was invented precisely so that the bats could carry enough incendiary punch. Whilst the bomb was never deployed in earnest, the designers did at least secure proof of its viability when some armed bats escaped during trials and flew into the eaves of the buildings of a US air base, which these suicide-bat-bombers then burned to the ground.[13]

Whilst such synanthropism is denied in policies that aim to separate bats from people, bats' curious place in mythologies and ontologies plays on lingering communicative possibilities—even ideas of human embodiment, mimesis and of shifting into bat ways of being and perspectives. Some of the creatures' habits and characteristics, such as hanging upside-down, and occupying a nocturnal world, play into a sense of bats as opposite to humans. A capacity to merge into bats' very different *umwelt* thus becomes a source of power for certain exceptional people, whether feared or revered. European literature and popular culture often represent bats as an 'other', with lives and powers very different from those of people; sometimes as unclean or allied with devils, malevolent spirits and witches that roam the land when darkness has fallen.[14] Shakespeare associated bats with spells and curses harmful to people and their social lives. Slavic folklore has, since the seventeenth and eighteenth centuries at least, associated bats with vampires: a framing elaborated through Bram

Stoker's *Dracula*, published in 1897, and the mass of fiction, film and imagery that followed. Another fiction valorises the 'human bat' as a superhero: as Batman. These instances are interesting as examples of diverse human narratives and genres that implicate bats, and more especially because they provide an opportunity to think of relations with non-human beings not simply through anthropomorphism, but in terms of hybridity, premised less on simple projection than on the mimesis that is at the heart of communicative endeavour.

Myths and stories that associate bats with witchcraft or vampire-like activity are also to be found even in parts of the pluriverse where more everyday framings dominate, coexisting alongside them. In Cameroon, for instance, people, bats, owls and bushcats can be said to be witch shapes, capable of sucking out the life of a sleeping person. Should a bat or an owl come near the house, or a bushcat defecate in the compound, the owner must go at once to a diviner to ascertain the remedy required. In Sierra Leone, the large hammer-headed fruit bat has been characterised as 'Boman', a creature that could turn into a stone or a snake at will and could suck the blood of sleeping children.[15] Vampire anxieties have provided a means to reflect on experiences of violence and exploitation, whether colonial or more contemporary, geopolitical or local.[16] In 2017 at least nine people were killed in rural southern Malawi, accused of being vampires.[17] Mythologies among American First Nations also draw on similar mimetic possibilities.

Human–Bat Viral Entanglements and the Structural Biosemiotics of Health

To turn to our third issue, a different set of framings of bats has come to pervade scientific and popular discourses, emphasising their role as hosts for viruses that can be harmful to humans and other animals. The association of bats with dangerous outbreaks of disease—epidemics and pandemics—has entered the public imagination, reinforced by popular media, in ways that can support the idea that human–bat encounters are rare and dangerous and that bats and humans should be kept apart. Problems with this framing are revealed by a structural-biosemiotic approach to human–bat communication, both in landscapes and, more suggestively, in relation to bodily viral interchange.

Scientific arguments concerning the origins of disease outbreaks of Ebola virus disease, Marburg, Nipah, Covid-19 and others have pinpointed the 'spillover' of viruses from bats to humans, whether direct or via other animal intermediate hosts.[18] Ebola epidemics have been attributed to direct human contact with bats, as in a dominant explanation of the origins of the 2013–16 West African outbreak that traces it to a toddler infected whilst playing in proximity to the bat roost in a tree in the village of Meliandou in Guinea's forest region. In other outbreaks, in East and Central Africa, transmission is supposed to have been from bats to primates and thence to people, via hunting and butchering.[19] Nipah virus disease, first identified in Malaysia in 1998, has been attributed to a bat-hosted paramyxovirus infecting pigs and then humans;[20] while a different viral strain in Bangladesh infected people when they were harvesting and drinking raw palm sap contaminated by bat secretions.[21] In Australia, Hendra virus was transmitted from fruit bats to racehorses and the vets who tended them. Many uncertainties surround the detail of such spillover pathways, as well as with regard to the balance between animal–human spillover and human–human transmission as the origins of particular disease outbreaks.[22] The key points here are that bats have become prime suspects as viral hosts, and key targets of the virus hunting in non-human nature that has become central to the contemporary sciences of pandemic prevention.

The proposed solutions that follow frequently focus on separating bats from humans. Environmentalists concerned with West African forests now find alliances with virologists in promoting conservation approaches that minimise human–bat encounter. Thus arguments for forest protection now include keeping people safe from viral bats, on the supposition that deforestation would bring greater contact.[23] Sanitary separation by culling or removing bats is also envisaged: in Australia, after fifty outbreaks of the Hendra virus killed horses, owners, trainers and vets, strong political lobbying from the horse industry called for a national cull of flying fox bats 'to check the southward spread of the lethal Hendra virus'.[24] Escalating outbreaks had damaged the multi-billion-dollar horse breeding and racing industry. Local councils were granted extra power to kill the bats and move bat colonies. Following several Marburg virus outbreaks in Uganda that have been associated with caves and the fruit bats inhabiting them, and which killed tourists as well as Ugandans, the Ugandan Ministry of Tourism instituted a cull in 2007. When foreign tourists began to cancel trips to Uganda during the more recent Ebola and Marburg outbreaks, the minister declared the intention to 'eliminate animals suspected to be carrying viruses of Ebola and Marburg', in an effort to safeguard the 10 per cent of GDP that tourism brings.[25] In many Latin American countries, since the 1970s, governments' answer to rabies outbreaks has also been to cull the vampire bats implicated in its transmission. In Ghana, when Henipah and then Ebola antibodies were found in bats, vets were concerned about revealing this to the public, especially as one colony was living above a public hospital: calls for preemptive culling would face them with a dilemma, as wildlife conservationists argued that bats should be protected because of their roles in biodiversity conservation.

These policy responses and the scientific and public imaginaries linked to them all suggest that human health is best served by separation from bats. Yet there are many questions regarding the feasibility and effectiveness of such separations—questions that are brought into sharper focus by viewing them through a structural-biosemiotic lens. First, most bats escape culls, and culling itself both raises the risk of human contact and disperses bat colonies, thus distributing the potential for spillover over a wider area. It can lead to an 'outbreak pulse', as the dispersed bats infect new roosts that have not recently been involved in active transmission, and where the virus sheds and spreads. The problems are not simply ecological, but also have communicative dimensions. Bat colonies codevelop with their viruses in emplaced ways—in what might be construed as interspecies bat–viral cultures—and dispersal mixes this up, creating cultural and viral exchanges, and increased possibilities for the appearance of new and potentially dangerous viral strains. Dispersing bat colonies also disrupts forms of emplaced communicative orderings amongst bats themselves, as well as with people with whom they cohabit, and the associated stress in bat colonies has been associated with increased viral shedding. Perversely, attempts at separation can thus increase the likelihood of human zoonotic infection.

Separation also proves unfeasible with regard to the human side of the disruption of more-than-human sociality. For people whose everyday lives are entangled with bats, such disruption can come over as inconceivable, disrespectful and dangerous: inconceivable and disrespectful because bats are a long-term presence, long-term cohabitants of landscapes, and the ordered, entwined companionship between humans and bats is seen as inevitable; dangerous, because the mutual accommodation of human and bat lives is central to the more-than-human ordering that ensures flourishing ecological assemblages and sociality, and is moreover emotional and affective. Such perceptions underlay the incredulity of Meliandou's villagers when visiting scientists told them that a toddler playing with a bat in a village tree was the cause of the Ebola outbreak that struck his family. For one thing, the villagers had a different explanation (that the disease came from an unwell woman seeking treatment in the village);[26] for another, the human–bat interactions that were being designated infective by outsiders were integral to long-term communicative orderings germane to healthy lives and ecologies as the villagers experienced them. In Bangladesh, meanwhile, in the context of Nipah virus, respect for cohabitation lay behind successful collaboration between researchers and villagers to develop a practical approach that maintained sharing of space while reducing the risk of viral transmission: the placing of locally made palm-frond 'hats' over vessels in which date palm sap was collected, so that the bats also seeking sap did not contaminate it.[27] Such more accommodative approaches to addressing zoonotic disease exemplify 'One Health' approaches, in which human, animal and ecosystem health are addressed in more integrative ways.[28] Appreciating the communicative orderings that codevelop amongst bats and people both lends weight to the advocacy of such approaches and can help in their elaboration.

A third arena in which attention to interspecies communication reveals species separation as problematic concerns more specifically bodily exchanges, playing out at an intra-body and microbial scale. Viruses that can infect humans are found more often amongst the animals with which humans cohabit, as entangled lives also, over time, foster repeated viral interchange. This is now recognised by virologists, who observe that 'virus transmission risk has been highest from animal species that have increased in abundance and even expanded their range by adapting to human-dominated landscapes', and thus '[d]omesticated species, primates and bats [have been] identified as having more zoonotic viruses than other species'.[29] Spillovers are most common from species with which we are in regular contact or with which we have coevolved in the past. This is because cross-species viral interchange works both ways. In multispecies, intertwined lives, not only do humans catch viruses from animals, but they also transmit them back again. Processes of viral interchange are an intrinsic part of our multispecies, companion relations. These processes build up over time to give certain viruses advantages in getting into and thriving within human cells. The result can be protective of as well as harmful to human health. Many of the viruses we live with as a result of such interchange do no harm, and some do good, squeezing out, or producing protective immunity against, other, more dangerous ones. For example, this appears to be one reason why Covid-19 was relatively less devastating in some African settings (the so-called 'African paradox') where people seem to have had prior exposure to related coronaviruses that conferred subsequent immunity to, or mitigated serious effects of, SARS-CoV2; exposure that researchers suggest is linked to long-standing viral interchange with bats.[30] Attempts to separate humans from viruses, and from the multispecies interactions that foster such viral interchange, could thus in fact undermine human health, by reducing the potential for such protection. Instead of what Jamie Lorimer refers to an as 'antibiotic' approach in the broadest sense—predicated on the separation of human bodies from the microorganisms with which they have codeveloped—a contrasting 'probiotic' view would recognise and appreciate the inevitable coexistence of human bodies with many microorganisms, including but not exclusively viruses, in ways that can on balance be helpful to humans, albeit occasionally harmful.[31]

These viral exchanges can be understood as communicative; as semiotic. Communication at the cellular level intersecting with genetic codes and immune systems is a major focus of biosemiotics inquiry: that is, the study of 'endobiosemiosis'. A structural-biosemiotic approach would emphasise how viral exchange indeed involves taking advantage, to an extent, of broader genetic coding that we have in common with naturekind, but also how our codevelopment with viruses and our more-than-human companions has resulted in more emplaced immune system assemblages. These can now be understood as the communicative outcome of previous viral exchanges, creating multispecies immunological cultures that in part account for the differences seen across human populations.

Human bodies (like bat bodies) are thus internally multispecies, in that each carries the traces and legacies (memory) of both current and prior interspecies viral exchanges. This exemplifies the broader point that, as biologists and anthropologists alike remind us,[32] the human body is a multispecies assemblage, comprising a plethora of microbial communities, many of which are crucial to metabolic processes that are central to our evolutionary and ecological becoming. Growing evidence shows how the internal microbiome affects human (and animal) ways of being, physical states and even character: for instance, the fifth of humanity, globally, that harbours the *Toxoplasma gondii* virus carried by cats is now understood to be more risk-taking and entrepreneurial as a result.[33] This growing evidence further underlines the unfeasability of separation, rooted in an antibiotic fiction that is at odds with the reality of our biotic selves within naturekind.

Across the pluriverse, human health is a more than human matter, involving multispecies interactions within and across human and non-human bodies and ecologies. This is variously elaborated, or played down. Arguably much of modern biomedicine, along with mainstream ecological sciences, seeks to separate—whether this is the separation of people and animals and ecosystems deemed helpful in preventing zoonotic diseases, or the antibiotic approaches that dominate therapy. Our brief foray into the interactions between humans, bats, and viral ecologies is suggestive of the limitations of these approaches, and their dangers. Perhaps medicine, and those who aspire to One Health policies that address human, animal and ecological health together, could learn from the everyday lives of those who live more obviously within more-than-human relations and conceive of these as inseparable. This might also help in building approaches that negotiate more effectively and sympathetically the delicate balances between protecting the health of humans and that of non-human beings and ecologies, all conceived as parts of naturekind.

8

Forests

Forests, whether tropical or temperate, highland or lowland, large expanse or small grove, offer particular opportunities, values and qualities of life, livelihood, identity and immersion. A basic definition of 'forest' could be 'an area of land dominated by trees', but there are myriad definitions in use around the world to assess whether something is or is not a forest, or divide forests into subtypes, based on tree height or density, land area, terrain, legal standing, ecological function and so on. The United Nations Food and Agriculture Organization (FAO), which claims some authority through its regular Global Forest Resources Assessment, defines a forest as 'land spanning more than 0.5 hectares with trees higher than 5 meters and a canopy cover of more than 10 percent', but only recently added an exclusion, that this 'does not include land that is predominantly under agricultural or urban use',[1] as until then many towns, cities and farms qualified as forest given that tree densities in streets, gardens and hedgerows met the stated thresholds. But over and above such formal definitions, forests (as opposed to plantations) are usually distinguished by their very density and diversity of tree, plant, animal and microorganism interactions, in multilayered, complex networks. Forest ecologists have been inclined to consider these as 'forest ecosystems', more or less intact, and to analyse the nutrient and energy flows and feedbacks among the plants, animals and fungi and linked to light, heat, water and soils that shape forest composition and dynamics. The flows in question are of material elements, between entities that are framed mechanistically, and not as communicative or conversational.

Here we consider forest assemblages through a structural-biosemiotic framework, from the Upper Guinean region of West Africa to the American Midwest and the South Downs of the UK, to probe how people engage with forests as part of communicative communities, picking up on their structured grammars and signs, packaged into metasigns and narratives. We hardly recognise this language on entering a forest, until we discern matter out of place: a rusty Coke can perhaps, as instantaneously as we would a misplaced word this sentence in. Such communicative relations with familiar forest assemblages are so emplaced that peoples' relations with them are through dialects that they codevelop, normalised subliminally in everyday interactions, though they still might notice an unusual bird, an oddly fallen tree, or an object stuck in one. And at the core of our arguments is that it is not just we humans who discern such matter out of place so grammatically, but that all creatures, attentive to the scents, sounds and movements around them, might do so. Indeed, it would be rash to suppose that an evolutionary outcome that enables us to perceive such matter out of place so instantaneously in forests just as in language should be restricted to humans, and so it would be equally rash to suppose that what is the case for physical perception in more-than-human worlds does not extend to communicative perception in such worlds, given that there is no clear distinction between these.

Encounters with forests provide an exemplar through which to evidence such reasoning, and in doing so, they allow us to consider the place of temporality and rhythms as part of linguistic orderings, whether in diurnal or seasonal cycles, or the traces of our past in the present; in living forests as legacies and 'archives' of more-than-human lives and communities.

An analysis that foregrounds communicative experience, as we then go on to show, helps us to understand the emotional, affective experiences that people have in and with forests, whether as part of forest-engaged lives or in more temporary forest immersions. This can cast new light on what it takes for people and forests to flourish together, for instance in aligning human purposes such as farming and food-gathering with the maintenance and resilience of wider forest life. Appreciating the significance of more-than-human communication in this way, and its emplaced quality, presents challenges to forest management and conservation approaches premised on modes such as commodification and offsetting that are premised on the erasure of such emplaced communicative relations.

Forests as Communicative Communities

Grounded in his experiences in Amazonia, anthropologist Eduardo Kohn's apprenticeship in the ways of Ruha people and their non-human cohabitants drew him to understand the flow of life in a forest as intensely communicative. In his *How Forests Think* he conveys a living world of the forest, its plants and animals, as consisting of sentient beings that are forever producing and responding to signs in relation to each other, to the ground below, and to the wind and rain that flows through them.[2] A tree, when it falls, conveys the sound of its fall, which will have significance for the beings nearby that sense it. A bird as it lands on a branch to feed from the fruit of a forest bush will rustle the leaves and move a branch, and other life forms—birds or insects perhaps—will pick up on this and respond. As the life in a forest perceives and responds to the signs of other forms of life, so a constant process of signing, reflection and action unfolds, with one sign making possible other signing processes. Forest life is 'constitutively semiotic', Kohn observes: entangled in sign processes, involving multiple modes—visual, chemical, aural and more. Those who inhabit or spend time in forests, humans and non-humans alike, are embroiled in this complex unfolding of meanings—conversations—with all the living world around them; in 'greater than human webs of semiosis'.[3]

Some ecologists, too, now reconceptualise what were once considered mechanistic flows and interactions in forests in more communicative terms. They reveal multimodal signalling processes amongst plants, animals and fungi: signalling that in forests takes place within, and helps to shape, dense interconnected assemblages. These networks of communication extend from trees and their canopies down into the dense fungal networks in the soils beneath.[4] They draw together multiple modes, from the scent and chemicals that are important to signalling between trees and plants, to the chemical and electrical signals involved in the so-called wood-wide web that interlinks roots with fungal hyphae, or the visual, tactile and chemical signals through which forest plants communicate with insects and bats, for instance in pollination: all of which, in the context of this chapter's interest in assemblages, we can now see as contributing to complex multispecies communicative networks. The density of such networks in forests certainly distinguishes them from young plantations where trees have yet to develop such interconnections. As Peter Wohlleben captures it, plants in such formations are 'isolated by their silence', and so are easy prey for insect pests.[5]

Forest ecologists such as Suzanne Simard and Peter Wohlleben even describe forests in linguistic terms, as 'speaking', just as Kohn considers them as 'thinking', and in ways that are more than just metaphorical or anthropomorphic. But what kinds of signing are envisaged in such forest semiosis? Kohn, in common with most biosemioticians, turns to the Peircean conception of iconic and indexical signs, construed as prelinguistic. He prioritises sound, drawing attention in particular to iconic signs—sounds which resemble what they signify—such as the 'splosh' of water or the 'errrrrr-bash' of a falling tree. Other signs highlighted by biologists and ecologists, as well as Kohn, such as direct chemical signals, are indexical, acquiring meaning through being part of what they are connected with. But none of these scientists consider how signing might work in symbolic ways. A structural-biosemiotic framework helps, however, by focusing on how meaning is created and exchanged in multispecies forest worlds not through isolated signing processes in which signs are unrelated to other signs, but through the ways signs are combined in grammars, however emergent, and in the wider packaged significations rooted in experience.

Considering the language of forest assemblages in this way suggests that in some respects, immersing ourselves in forests (and indeed other kinds of assemblage) can be seen as a multispecies version of what anthropologists or others do when they settle in a new place. As people immerse themselves in a new social world they experience with surprise the contrasting social grammars that become apparent to them, but which are subliminal to those they now live among: as subliminal as their spoken grammars are to the way they speak.

To consider forest assemblages and people's engagement with them through a structural-biosemiotic framework, we can begin in a field in foothills of the UK's South Downs where we authors started to plant trees a few years ago. We chose a diversity of local woodland species, but—newly attuned to emerging ecological sciences—we were aware of their vulnerabilities, as the trees would be establishing themselves in isolation, yet encouraged by the possibility that, with soils and undergrowth having long been left uncultivated, they might develop new, if unknown, communicative networks and the resiliences these might afford. Over the ensuing years the oak, hawthorn, beech, hazel and other saplings have grown, the undergrowth has thickened, and numerous birds and insects have become inhabitants or regular visitors. The place is gradually becoming a woodland, a forest, and with it, a nascent, dense web of semiosis. We cannot pick up directly on much of the signalling between the codeveloping tree, plant, animal and soil life, which goes on out of view underground, when we are not present, or through sensory modes unavailable to us as humans. We do manage to pick up on the changing quality of this patch in more structural ways, however, as the vegetation becomes dense and tall (in contrast with the adjacent grassland in which it was established) and develops sounds: buzzes, clicks and other sonic signals (in contrast with the more silent grasses, especially when mown for hay). Density and chatter—the words being inadequate shorthands in our human language—are thus not just epiphenomena that package all the constituent signing processes, but together become a metasign making sense in relation to other visual and soundscapes, with a distinct 'feel' of being in a nascent forest. This nascent forest nevertheless feels very different from (and acquires its sense in relation to) the patches of longer-established woodland in the valley bottoms a mile or so away, where other metasigns—darkness as opposed to surrounding light; distinctive, more musty smells of rotting vegetation; softer earth underfoot—combine to convey an older forest with denser communicative networks.

As in this example, it is the way signs juxtapose and combine, picked up as a package in relation to other packages, that adds up to the feeling and experience—the meaning—of being in a forest, and not in some other kind of landscape; of being in 'forest' as distinct from all that is 'not forest': not open grassland, not urban land, not farmed surroundings, not farmland or a regimented plantation. A great deal more is communicated by the signs linked to visual and aural senses: a sense of distance derives from the combinations of geometry, parallax, colour, planes of light, definition and texture that artists since the time of da Vinci have experimented with and that shape perception of perspective. Each sign acquires sense in relation to its particular field of meaning, but integrates subliminally with other indicators of perspective, as well as with the equally complex sonics and scents of space, to become packaged into equally subliminal metasigns involved in the language of distance. Experiences include the combination of darkness and shadiness, registered visually; of coolness, of scents and smells—fruity, earthy, upwelling on a warm day, conveying decomposition. The signs we pick up on include the sounds of forest creatures and of rustling leaves and branches; a chatter that surrounds us even if not addressed to us. In short, one 'feels' a forest; but if we speak of such experience as 'embodied', this is only as a shorthand to capture the subliminal way we package familiar combinations of sensorial signs into hierarchies of metasigns, as oblivious to the myriad sign fields and grammars involved as we are to the equivalents in our human spoken language; oblivious, that is, until we accidentally knock our head on a branch, or hear an unfamiliar growl.

The forest is felt and experienced differently depending on its own specifics (of ecology and more), and on who one is, but however it is experienced, this is nevertheless through a structured combination of signs and hierarchies of packaging all making sense in relation to experience of a structured field of similar packages; in relation to our sensorial language codeveloped with the places we know. The forest we are in acquires meaning, and is felt, by way of contrast: as being different from a non-forest (and other kinds of forest), and also from a plantation with its ordered rows of trees.

Our South Downs experience also highlights to us how a forest assemblage can be experienced as a more-than-human, 'communicating community', as it were. People who know a particular forest enter therein a communicative world that has become familiar and intimate; the familiarity comes from experience, from codeveloping a language with the forest, through which to interpret (read) its signs and metasigns. We have come to feel this in the forest patch that is codeveloping with our care, and as we offer the trees small signs of encouragement, such as some extra water in prolonged dry periods and weeding around, it has come to feel like an expansion of our circle of other acquaintances. Just as people have circles of friends and neighbours, so communicative encounter with a forest can expand the sense of friendship and friendliness out into wider and non-human circles. And just as one cannot know the detailed meanings of every communicative act, interpret the thoughts or foresee the intentions of every person in a social setting, so in a forest we act on what we can pick up on, and proceed carefully around what we cannot.

In the accounts of Robin Kimmerer, the pecan groves that occupy lowland and streamside areas threaded across the prairies of the American Midwest offer a powerful example of how the interactions and dynamics in forest assemblages can be conceived when viewed through a communicative lens, and in structural ways.[6] As a biologist and member of the Potawatomi Nation, she describes how the pecan trees all fruit together at unpredictable intervals every few years, 'a boom and bust cycle known as mast fruiting', with which both people and squirrels align in a 'feast and famine' harvesting practice. Many forest ecologists and evolutionary biologists interpret this synchrony mechanistically: as the simple outcome of each individual tree's energy flows, or in terms of a predator satiation hypothesis, configured in a dynamic between trees, squirrels and their predators, according to which, when the trees produce more than the squirrels can eat, squirrel numbers increase, along with populations of squirrel predators (foxes and hawks), and then, when through starvation and predation the squirrel population plummets, the trees fruit again. Yet Kimmerer interprets this cycle in more communicative ways, suggesting that as squirrel populations fall,

> the woods grow quiet without their chattering. You can imagine the trees whispering to each other at this point, 'There are just a few squirrels left. Wouldn't this be a good time to make some nuts?' All across the landscape, out come the pecan flowers poised to become a bumper crop again. Together, the trees survive, and thrive.[7]

Expressed anthropomorphically, Kimmerer's account draws attention to the potential signing processes involved in the predation cycle. She notes, however, how even biologists continue to puzzle over the trees' collective action and its timing, finding the predation explanation inadequate:

> The pecan trees and their kin show a capacity for concerted action, for unity of purpose that transcends the individual trees. They ensure somehow that all stand together and survive. There is some evidence that certain cues from the environment may trigger fruiting, like a particularly wet spring or a long growing season [...]. But, given the individual differences in habitat, it seems unlikely that environment alone could be the key to synchrony.[8]

Instead, the puzzle is resolved by turning to communication between pecan trees as the key to synchronisation, either through pheromone signals, or through fungal networks that redistribute carbohydrates from tree to tree and through which 'they weave a web of reciprocity, of giving and taking'. In this move from a mechanistic to a communicative conceptualisation of pecan dynamics, there is an alignment between the emerging science of forest biosemiosis and Indigenous theorising. As Kimmerer puts it,

> In the old times, our elders say, the trees talked to each other. They'd stand in their own council and craft a plan. But scientists decided long ago that plants were deaf and mute, locked in isolation without communication. The possibility of conversation was summarily dismissed [. . .]. Yet there is now compelling evidence that the trees *are* talking to each other.[9]

People and animals become part of the multispecies conversation involved in mast-fruiting and the collection of pecan nuts, and this enables a mutual flourishing in which 'soil, fungus, tree, squirrel, boy—all are the beneficiaries of reciprocity'. Although Kimmerer does not elaborate in these terms, a grammatical ordering can be discerned in such synchrony, in which multimodal signing processes amongst trees, fungi, squirrels and humans are coordinated with each other and over time such as to sustain a 'feast and famine' ordering in more-than-human life that is important to trees, squirrels and humans alike. Where Kimmerer does elaborate further is in conceiving of pecan groves and other forest assemblages in the prairies as a community of communicating kindred. She describes how, '[w]hen I am in the woods with my students, teaching them the gifts of plants [. . .] [a]lthough they still have to learn scientific roles and Latin names, I hope I am also teaching them to know the world as a neighbourhood of nonhuman residents'.[10]

On the northern fringes of Guinea's forest zone, islands of semi-deciduous rainforest are often to be found in rings that surround each village. Living in these, we found them to be experienced in similar, intimate, conversational ways by our Kissi- and Kuranko-speaking hosts, from whom we gradually picked up conversational possibilities. In their experience, these patches of forest emerge around settled human social life.[11] A peri-village forest grows quickly over the years through a combination of intentional planting, encouragement, and as an inevitable outcome of everyday practices, as people garden, tether animals, collect grasses and suppress fire, and more—in addition to the activities of animals and birds that frequent the nascent forest. These forests become established as islands in a sea of savanna, and old forests formed in this way still stand on the sites of villages long abandoned by their human inhabitants, but forming distinctive parts of the mosaic of vegetation in landscapes of forest, savanna, farmland, rocky hillocks and marshy river bottoms. These forests are places of practical encounter, but also of communication involving plants, animals, and other beings, and on many walks through, we attuned ourselves to become part of the conversation. Thus, walking along a forest path, we pass trees and an understorey whose leaf shapes and smell signal their identity as being edible for birds, monkeys or people, picking some later to make a sauce for cooking or to brew up as a medicine. We enter a web of connection in which other animals are also picking in their own ways, to which the plants respond. We come upon a chain of soldier ants crossing the path, carrying their burdens of leaf fragments, a sign of their connected place in the web of forest soil–plant relations but also signalling potential pain, if they bite you; so we avoid their chain to pass unbitten, leaving it to re-form or perhaps go off in a different direction, encountering other parts of the forest floor. The chatter of calling birds, falling branches, rustling leaves is everywhere around, communicating signs to other animals and plants which they pick up on and respond to. All this is the communicative conversation of the forest, as sounds, movements, smells, tastes convey things that are picked up by those nearby. Lamin, who helped translate the forest world to us as we became familiar with the forest language so familiar to him, would describe the different trees, knowing not just their names and their usefulness to people but the ways they would attract other animals for their seeds or fruit, and their habits of growth, fast or 'little by little by little'. Manty, similarly, described the leaves in the forest behind her kitchen as falling into the stream in the dry season with a 'dukudukuduku' sound. Other beings in the water sense the leaves that fall in, leading to their decay into fragments that fish would then eat (but which, as Manty complained, 'spoiled' the water for people to drink). We could go on; the point is that attentiveness, interactions and close experience reveal the forest as a multi-meaning, multiply communicative world, in which we who enter it inevitably become an intercommunicating part. Lamin, Manty and others in the region's villages lived this daily. Multiple signing processes amongst humans, plants, animals and fungi are integral to the creation and sustaining of the forest island, to shaping the patches of forest of different ages in savanna, and to the meanings these then convey to those living in them and nearby. This is an ordering deeply entangled with the ordering of human social life and livelihoods, and the practices of residence, movement, gathering, farming and associated social relations they involve.

We draw on these examples from the UK, North America and West Africa to suggest that forest assemblages can be framed and experienced as communicative more-than-human communities, codeveloped and changing through engagements between their many inhabitants in semiotic ways. Such a framing might be seen as a different way of describing what forest ecologists are accustomed to call 'forest ecosystems', or 'socio-ecological systems', to recognise the roles of people in them too; for what is a system if not a structured ordering of elements? For much of the twentieth century, conceptualisation of these dynamics placed an emphasis on a more or less linear vegetation succession towards what was referred to as the 'climax' vegetation that a given place, with its particular climate or environmental conditions, could and would support. This idea that each place would acquire a particular climactic vegetation if left alone has since been superseded by conceptions of forest communities as emergent, non-linear, complex adaptive systems. Accordingly, older ideas that humans only disturbed the 'natural' forests of a particular place, and created 'anthropogenic subclimaxes', have given way to more open appreciation of more diverse, multi-way interactions between people and forest systems. Yet either way, these still remain mechanistic ecosystem framings and sit at odds both with the growing ecological revolution that considers forests in terms of communicative networks, albeit, as yet, prelinguistic ones, and with a conception informed by structural-biosemiotic analysis that would appreciate the orderliness captured in the different ecosystem framings, but treat them in terms of syntax and narrative.

Contrasts between the dominant framings of forest ecological science and more communicative ones also turn on questions of what kinds of beings we consider to inhabit forests. Who or what are part of forest communicative communities? In particular, in some quarters of the pluriverse, non-human forest inhabitants are understood and treated explicitly as kin. Thus amongst some American First Nations peoples, such as the Potawatomi Nation to which Kimmerer belongs, the sense of forest as a neighbourhood beyond the human is underpinned by ontologies in which forest trees are considered as persons, and as subjects with particular social places in more-than-human worlds.[12] Many of the Ruha people whom Kohn lived with, as reported across Amazonia, would understand their communicative forest world as one in which humans, animals and plants are all part of a single more-than-human culture, but that take many different natural forms within it, which can occupy shifting positions in reciprocal predator–prey relationships.[13] In West Africa, multispecies forest worlds are frequently also home to entities such as *jina*, spiritual beings some of which are often understood as being the region's original inhabitants, and with whom people had to negotiate when they arrived and wished to settle. When walking through patches of forest near Kissidougou, our human companions would sometimes point out the particular rocks or trees that *jina* inhabit. They can be capricious, and those who do not know where they live are warned to take care and not to wander off paths, so as not to disturb their houses. The notion of part of a forest as a '*jina* house' (*jina so* in Kuranko) does not just allude to a kind of individual being that would never appear in forest ecosystem analyses, but actually in itself constitutes an alternative category for understanding such 'ecosystems'.

The more-than-human inhabitants of forests can include all sorts of beings: plant, animal, spirit and more. Depending on the framing in play, these might be attributed different kinds of status—as animate or inanimate, and as having agency and intention, or not. Agency is notoriously hard to pin down, and all the more so in an assemblage where change might be attributed to collective interactions as much as to an individual being—something we discussed in relation to bees in chapter 5. In one framing, agency might be attributed to a 'council of trees'; in another, to a 'system', or to a 'spirit'. Yet people do not need to understand non-human forest beings as agentive to appreciate forests as communicative and kindred. We authors certainly do not, in our South Downs forest patch, and nor did our Kuranko and Kissi human companions and guides in Guinea, for whom personhood may have been attributed to *jina* spirits, but generally not to trees and plants. Instead, the sense of kindred, of neighbourliness, seems to be borne more of a general familiarity and intimacy, in which one is aware of multiple signing processes going on between all forest beings, of entering and immersing as part of such a world, and of being able to read some, if not all, the specific signs.

Emplaced Forests

The grammars and narrative forms we codevelop with forests are thoroughly emplaced, as forest assemblages constitute places in and of themselves and also sit in particular spatial contexts. Thus a patch of forest might sit in a field or savanna; on a mountain or by a river; in or adjacent to a city. Physical emplacement clearly shapes what goes on within, as slope, winds, scale and boundary effects influence what organisms thrive and how they relate to each other. Yet although everything is emplaced, and very particularly so, there is still something about forests that makes them feel familiar if one knows another one that is similar. When we enter a different part of the forest, where the soil, slope and species vary slightly, the broader signs and orderings, or grammars, through which we converse still hold. We still know the language: it still speaks to us in the particular 'forest' dialect we have worked out with it.

If one thinks of a familiar forest as communicating correctly with us, this opens up the possibility of there being something awry; of matter out of place screaming significance. Matter out of place might signal danger, but might also reframe the narrative genre (metasyntax) in which metasigns acquire meaning. Coming upon the tracks of a leopard, for instance, might not simply be out of place, but switch the narrative genre in which we were initially conversing (perhaps 'climax forest ecosystem' or 'complex adaptive system', or indeed, 'rom-com' or equivalent) to 'thriller'. The forest ecologist who thought they knew the forest in a mechanistic way, on seeing leopard spore now feels scared, realising with a jolt that the sounds of disturbed birds and broken branches, and a musty feline scent have new meaning. Whereas the former narrative structure distances the ecologist from the forest they are in, and removes them from an emotional relation (or more correctly, configures them in a relation of emotional indifference), the latter narrative places them firmly within the communicative forest, and within an emotional entanglement of fear. Or one might come upon a clearing where the tyre marks and timber stacks suggest loggers have been; or a path signalling that there has been industrial development. Such stacks of planks beside a forest path as one walks through a Guinean forest island, or the clearings where ash trees have been felled in in the forests of England's South Downs, can be evidence of loggers' disruptive interventions in a natural world, if that is the organising narrative (metasyntax) of encounter. For forest managers the narrative might concern addressing a fungal disease. And for the chainsaw operator—or the person who has invited them in—the narrative framing would be rather different again. Yet for those who perceive the metasign of the stack in relation to narrative genre (metasyntax) of ecosystem dynamics, or in relation to *jina* spirits, the stack becomes matter out of place, noticed and with the potential to connect productively with an ecological politics that attempts to counter or mitigate disruption.

The emplaced ordering of a forest assemblage can sit within wider landscapes which are themselves ordered. In Kissidougou, for instance, forest islands are part of a mosaic landscape that also includes areas of grassy savanna, and areas under varying stages of regeneration according to cycles of bush-fallow farming. The distribution of forest, savanna, fallow and field in the landscape shifts, over annual seasonal cycles, over the five- to ten- year cycles of bush-fallowing, and over the decades- and centuries-long timescales involved in the birth of forest islands around new villages and their persistence on the sites of abandoned ones before decay. For the region's inhabitants, this temporal rhythm of the forest-savanna mosaic landscape is an aspect of life that is taken for granted. It relates to a wider landscape ordering that is mutually constituted with the ordering of human society, in human kinship and in the other social and political relations that alter where people live and how livelihood practices interlock.[14] The point here is that, by observing it through a structural-biosemiotic lens, we can appreciate this wider forest-savanna mosaic experience as creating the narrative framing through which signs (metasigns) in the landscape are communicative—in which meanings are conveyed through their juxtapositions and relation to the cyclical rhythms of the forest, savanna, field and fallow patches. Such meanings of vegetation, for Kissi and Kuranko villagers, include those pertaining to the thriving of settled social life in general, and to the flourishing of particular villages and families who trace their ancestry to particular forests. Human dialogues about these things are not separate from the conversations with the forest assemblages themselves.

Forest assemblages and the wider landscapes they are part of can also communicate meanings that convey their history, open to 'reading' as such by those prepared to look and listen. In Kissidougou's forest islands, people would sometimes point out the slightly different appearance and qualities of of certain vegetation, indicative of an old field or ancestral grave it overlies. They would point out where small groves had developed from the kola trees linked to particular families, or the vast silk-cotton trees that had grown up from sticks used for fencing by a long dead ancestor, whether to defend gardens from animals or villages from enemies in times of war. They would talk about the family histories that had led to the founding of the village where a forest patch had emerged, and the village decisions to keep parts of it intact, out of respect for ancestors, while allowing farmers and tree fellers to convert other parts for human purposes. Such forest island histories, and the local settlement and political histories enwrapped with them, are also related more formally in the narratives of elders. Yet whereas one might (as we once did ourselves) analyse these as human readings of a forest landscape and history, or as human narratives or 'oral histories', by applying a more-than-human and structural-biosemiotic lens we can appreciate the forest as a communicant in such narratives too. The communicative grammars of a present forest—its current multispecies world—actually reveals something of—speaks of—its complex past. The forest can thus tell its own story to those who have codeveloped a language with it. It does so through the multiple signs in its constituent entities, such as the existence of a particular tree or dense patch over a grave; the signs conveyed by its character and quality, such as being sparse or dense, intact or gappy; as well as by its position in a wider landscape. People who grow up with and live with forests, or are open to learning from them, are able to pick up on the meaning of its signs, having acquired this language in just the same way as they have acquired their human language; and they can hear (and are able to retell) the forest's narratives. It would be no different for other forms of forest life, albeit the language codeveloped by a bird, for example, would have its own particularities.

In contrast, those who have not acquired this language might be unable to read the landscape and its signs, or end up 'misreading' them, interpreting them on the basis of their own preconceptions or from framings shaped outside the experience of more-than-human ordering. We encountered many forest ecologists, scientists and policymakers who interpreted the forest-savanna mosaic in Kissidougou, and indeed across many other parts of West Africa's Upper Guinean zone, as the result of loss of a once extant forest; forest islands were for them relics, metasigns that acquired meaning in their juxtaposition to savannas in metasyntax (narratives) of decline—of destruction at the hands of the region's farmers. Their narratives, of a landscape 'half empty and emptying' of forest, contrasted absolutely with villagers' narratives of a landscape 'half full and filling', and of forest islands not as relics, but as the counterpart of settled social life. Foresters themselves can learn to read the metasigns differently, as experience in the forest radically shifts their metasyntax. A colonial forester, D. R. Rosevear, once admitted as much on returning to a high forest that he had visited thirty years before in Nigeria:

> What I had in my inexperience looked upon as glorious virgin growth, dating from the Flood, quickly revealed itself to my better experienced and disappointed eye as nothing more than secondary growth of moderately good quality. [...] But there was one curiosity. The abnormally large trees which had so impressed me in 1924 were still there, scattered throughout the forest in sufficient numbers to attract puzzled attention. I was still more perplexed when I discovered that they comprised two—and only two—species, the Sasswood, *Erythrophleum ivorense* and the Inoit Nut, *Pogaoleosa*. Why these two species? And why this young—in tropical terms—forest, indicating previous widespread destruction in a so markedly underpopulated area? And then it dawned on me. *The forest made it clear to me, like reading a book*, that the entire region had once been heavily populated, so densely in fact that the whole, except perhaps for the more inaccessible upper portions of the hills (which I myself never visited) had been intensely farmed, leaving no surplus area of undestroyed forest. What had become of this population; and why the two untouched species, obviously carefully preserved hang-overs from the original forest?[15]

Rosevear learned to read the signs within a new narrative—a new metasyntax—of the forest, assisted by attentiveness to the communicativeness not just of the people who live with forests, but also of their non-human inhabitants, and the assemblages of which they are a part.

Forest Immersions and Affect

We have been discussing forest assemblages and their characteristics as communicative communities from the perspective of those who inhabit them, often intimately and over long periods. For many, however, forests are far less familiar, and yet satisfy desires to be in the unfamiliar, enabling a switch and a difference from habitual daily worlds. Appreciating forests as communicative helps explain what happens when those who do not live with forests day to day—urban dwellers in the south of the UK, for instance—seek immersion in a forest. In what is usually thought of as 'getting close to nature', or 'reconnecting', one is entering a conversation, learning a language, developing a familiarity. This helps us to understand the appeal of actively promoted 'forest immersions', and how and why they have their effect. In this practice, those less familiar with forests visit them, or even take part in activities such as the 'forest baths' that are now advertised locally to us in Sussex.[16] To the extent that this 'works', and has power, one might suggest that it is because participants 'bathe' in a sentient, communicative world in which the forest and its communicative webs are working through them. And this changes people—in ways that, depending on who they are, and what they seek, they might experience as calming, grounding, reorienting, energising, enlivening, or more. At one level we might see and describe immersive experiences as about 'losing oneself', or 'escaping', and perhaps doing so with other humans, as a way of 're-grouping'. Yet the inverse is also possible—because through a process of co-becoming we are changed, momentarily at least. When people describe forest immersion as a 'grounding' or 're-rooting' of themselves, the process is perhaps one of recognising the more-than-human communicating self. Immersion can of course happen, but not 'work': we may return from our immersive experience and feel somehow unfulfilled, experiencing the same lack of engagement as tourists feel in social worlds whose language they do not speak. Is it that curated immersive experiences work precisely because, and only if, they are social and communicative, engaging with a lively and communicative place? This is what makes us feel good, or refreshed, or the many kinds of things one feels when on a walk or sitting in a forest. Whatever the precise emotion, there is affect.

Immersion in emplaced forests with layered landscape histories can be a way to connect, communicatively, with a longer-lived sense of our own human society as well, and of a multispecies social world. Forests archive human history. We can return again here to the example of the pecan groves in the US Midwest that Kimmerer describes, which occupy a landscape in which First Nations people have experienced histories of subjugation and violence. Over the past century the enforced movement of human populations onto reservations and into individualised forms of land tenure disrupted many aspects of people's entanglements with land and its non-human inhabitants. As Kimmerer recounts, those pecan groves that remain standing have come to offer forms of reciprocity and mutuality that have been lost in many other areas of life: 'Living by the precepts of the Honorable Harvest—to take only what is given, to use it well, to be grateful for the gift, and to reciprocate the gift—is easy in a pecan grove. We reciprocate the gift by taking care of the grove, protecting it from harm, planting seeds so that new groves will shade the prairie and feed the squirrels.' The Potawatomi Gathering of Nations is often synchronised with mast fruiting, and this is important not only materially, to feed the gathering, but also symbolically: 'As a nation, we are beginning to follow the guidance of the pecans by standing together for the benefit of all. We are remembering what they said, that all flourishing is mutual.'[17]

Communication and Alignment in Human–Forest Flourishing

As this example suggests, people who seek to work or engage with forests for human purposes, whether in securing food or livelihoods, or sustaining human social relations, may be able to do so better if they are attuned to what, and how, a forest is communicating. While some people have the luxury of being able to spend time in forests for pleasure or leisure, many need to work with them for their livelihoods. Those seeking to farm fallows, or cultivate useful trees under the canopy, or collect medicines or fruits, or make a hideout, are not just in the forest, but interacting with it for human purposes. Yet outcomes are better when activities are carried out in alignment with the communicative networks already in operation. Thus the siting of an encampment, for example, is more likely to be effective and comfortable if it avoids the ant trail through which these insects move their food, or the tree that is about to fall. Successful farming needs to work with the communicative interactions of tree growth and soil enrichment. This is what farmers do year on year in bush-fallowing, choosing sites, felling trees and bushes and burning to clear vegetation and create ash that enriches not just the crop, but also the soil for subsequent tree regrowth. These are human purposes, human projects, but they interact with a whole range of other 'projects': what the butterflies are doing, or the trees, or tree–fungal interactions, and so on, with or without intentionality. Things tend to go better when their respective purposes align, and rather badly when they conflict. Those who live with and from forests—hunters, gatherers, bush-fallow farmers—know this well. Whilst one might think of their practices as 'hitching a ride' on non-human nature—going with its flow—this is to give an overly one-sided impression of what is in fact more mutual: interactions between multiple human and more-than-human projects. And these are necessarily communicative: one needs to discern what the other puposes are, so as to align better with them, and so that each 'participant' can adapt to the others. Those gathering fruits read when the trees start to bear, and also the signs that birds might be after them. Aligned fruit gathering, for instance, might involve taking some for humans, and leaving some for the birds to feed on, and for these birds to scatter the seeds, favouring future tree growth.

Shifting cultivation, or bush-fallowing, as practised by the Kissi farmers we lived with in Guinea, illustrates this reading of signs and alignment of projects well. In the mosaic of forest, savanna and vegetation patches at various stages of farming and regrowth, farmers read the vegetation and soil signs which speak of whether an area has grown sufficiently as fallow to be ready to clear. Farmers clear land with an understanding of the temporal rhythm, a cycle, in which the felled forest assemblage recovers after a decade or so. Those igniting felled vegetation successfully may be cheerful, happy that a burn was 'really good', and whilst in part that emotion reflects the likely productivity of the rice crop that year, in part it is a matter of reading that the way the burn has happened—the visual and sound signs of a vigorous fire, the remaining black ash and charred stumps—will allow the fallow to recover. Those who work well with forest ecologies know how to interpret the signs so as to maintain the resilience of the cycle. In this way bush-fallowing is far from being the mere 'slash and burn' it is sometimes labelled as—and certainly corresponds still less to the worse labels, such as 'acts of vandalism', that have been applied to it in this region as elsewhere. It would be vandalism if it were simply a destructive act. It is anything but, however: it involves a working with the dynamics of a communicative world where farmers have a sense of the assemblage and its capacities, such that the purposes of all the different organisms and elements present can be aligned. One does not need to know the details of all the communicative networks in play—the messages of all the particular fungi, beetles, trees and undergrowth plants. These can be of interest: sometimes a farmer will indicate, for instance, why a specific tree indicates the age of the fallow, or where a termite mound suggests the fertility of a particular patch of soil. But generally, farmers practising shifting cultivation pick up on the higher order package of meanings communicated by the bush or fallow field, with the height or quality of vegetation registered as a metasign. Deep experience and intimacy give a better grasp of the details within the broad structure, enabling one better to read the subtleties: how old this fallow is; what its soil is like; how significant it is to particular families; and then one can align one's projects better. Forest clearing could be vandalism if it were simply a destructive act with no regard for the consequences. And bush-fallowing is certainly very different from clear-felling forest vegetation to make way for a novel human purpose in using the land—an oil palm plantation or pasture perhaps, or housing for urban expansion. These are not 'aligned purposes'; they are human purposes overtaking those of the forest, perhaps removing or completely replacing it. The result then is a shift away from the grammatical structure of a forest, towards something else entirely—a cityscape, a plantation, a grassland—with its own grammar, to be sure, but one very different from the forest's.

A structural-biosemiotic perspective also highlights the problems not just in understanding forests as uncommunicative, but in considering communication only in terms of separate signs (as Peircian biosemiotics does). Such a take is compatible with approaches to forest management and conservation that would split them into components, or separate processes, and with their commoditisation—as ecosystem services: tradeable biodiversity or carbon, for instance. The conflicts that have often emerged when forest assemblages are enclosed and sold for such purposes are to be understood in part as politico-economic: as conflicting with the established values, rights and livelihoods of human forest inhabitants, and resented as such.[18] Yet our discussion in this chapter points also to the ways in which such interventions sever both the more-than-human assemblage and the communicative networks that constitute it, and that enable people to live and thrive with forests in more convivial, mutually appreciative ways.

9
Soils

Soils are now high on the list of global environmental concerns; a crucial but finite, non-renewable resource under devastating pressures from erosion, degradation, poor management and loss to urbanisation and infrastructure development. Specialists raise the alarm that we are reaching or have surpassed 'peak soil', along with 'peak nitrogen' and 'peak phosphorous', losing even the chemical fertilisers deemed necessary to stave off or redress soil fertility decline. Others highlight how prevailing use of fungicides and insecticides damages soil ecosystems and contributes to soil degradation. The loss and exhaustion of fertile soils and the implications for our ability to feed ourselves, let alone to sustain the myriad other 'ecosystem services' that depend on soils, including carbon sequestration, is a focal topic of scientific and policy concern and debate. Alternatives to agro-industrial soil management have long been promoted by organic farming movements, and are now envisaged too in a range of other approaches, including regenerative agriculture, agroecology and permaculture, and in the appreciation of local and Indigenous approaches to soil use and care. Long-standing anthropological research on 'ethnopedology' considers how cultivators across the pluriverse classify and categorise soils and understand the processes that enable them to thrive.[1]

In this chapter, we consider how existing debates about human–soil relations and how soils might thrive underplay communicative dimensions. Some soil scientists and soil managers treat soils less as living than as a mere substrate or matrix in which crops are to be encouraged to grow. Others—soil ecologists—treat soils as living: as assemblages of a wide variety of organisms, from micro- and macrofauna to plant roots and fungi, and probe the interactions of these, as being crucial to soil quality. Yet the latter group generally conceives of these interactions as mechanistic. Our aim here is to reconceive living soils as communicative communities, sharing meaning amongst their biotic and abiotic components, and most particularly to explore how people participate in these processes, becoming part of more-than-human soil conversations—conversations that take place over overlapping timescales.

Conceiving of how people and soils are entangled through language beyond the human helps build on, but goes beyond, the rather static classificatory approaches that prevail in existing ethnopedological inquiry. It adds to agendas emerging from the social sciences and humanities that do consider living soils as more-than-human, emphasising 'relational materialities' that 'acknowledge symmetrically the emergent biophysical agency of soil ecosystems, their sociocultural constitution, and the dynamic interactions between those factors [. . . ,] an understanding of soils as dynamic ecologies in the becoming of which human beings are implicated, with whom they are shaped, and on which they depend'[2]—yet give little attention to communication as a component of such relations. Moreover, the approach advocated here can serve to augment anthropological and Indigenous scholarship and activism that appreciates how people have ties to their land. It allows us to perceive how soil and human processes become imbricated in each other through shared language and meaning; how their co-becoming unfolds in temporally and spatially structured ways, and how this can create communities and affective relations beyond the human that are important to mutual flourishing.

Soil Assemblages as Communicative Communities

Though sometimes considered as simply an essential background and substrate on which so much plant, animal and human life depends, soils are themselves increasingly recognised as ecosystems in which a vast array of living and decaying organisms are mutually entangled such that the whole is more than the sum of its parts. Rapprochements between ecological and soil studies add rich insights into the sheer quantity, variety and significances of soil macro- and microorganisms, fungi and bacteria.[3] Soil structure, texture and mineral composition depend on the ways macrofauna such as earthworms or termites transform and transport mineral particles, altering clays and their mix with other minerals and plant remains and developing organic humus. Nutrients on which plants thrive are not simply present 'naturally', adhering perhaps to the soil matrix, but are gathered and concentrated by bacteria and move through mycorrhizal fungal networks that themselves associate with plants. Plants release chemicals into the soil as sugars from photosynthesis and complex compounds, and other organisms these feed assemble and unlock many of the nutrients on which plants depend, via processes such as the 'fixing' of nitrogen from the air into nitrate and ammonium that plants require in their growth.

As knowledge of such interactions multiplies and as their significance for soil quality is increasingly appreciated, soils are exposed, in the words of philosopher Maria Puig de la Bellacasa, 'as a living world rather than a mere receptacle and input for crop nutrition.'[4] As she notes, notions of a living soil, once associated with 'organic' and radical visions of agriculture, are increasingly mainstream in soil ecology and soil science. This is not to say that all earlier soil science conceived of soils simply as inert matter—there was always interest in lively physicochemical processes and interactions, and soil biology has been part of soil science since its origins, with Darwin's appreciation of soil macrofauna being a case in point. Yet a paradigmatic change can be discerned in the increased significance of biota, from microbes to macrofauna, plant roots and fungi, in the very definition of soil. Organisms now *are* soil: 'A lively soil can only exist with and through a multispecies community of biota that makes it, that contributes to its creation.'[5]

For the most part, however, those appreciating soils as multispecies, living assemblages still understand them in mechanomorphic ways, and most scientists conceive of soil ecosystems as being maintained through instinctive and mechanistic interactions amongst their component parts. Foodweb models have become prominent to describe the incredibly complex interactions between species that allow the circulation of nutrients and energy. Emphasis is on 'trophic' relations amongst soil foodweb species that can include algae, bacteria, fungi, protozoa, nematodes, arthropods, earthworms, larger animals such as rabbits, and of course plants. The growing understanding of soil interactions then adds to the range of biotic and abiotic beings understood as parts of such webs. This increases appreciation of the dynamics of these interactions, leading soil systems to be portrayed as ever more multilayered, complex and emergent; but they are still seen as systems in which their multispecies components are driven by instinctual or mechanical response: interacting, but not communicating.

Observed through a biosemiotic lens, however, it becomes evident that soil assemblages are indeed communicative. Many of the interactions now highlighted by soil ecologists can be considered as semiotic processes through which soil organisms share multimodal signs and others respond to them. Soils have not received much attention in the academic biosemiotics literature, although, without using the term, a number of researchers and commentators now convey the notion that soil inhabitants engage in 'signing'. Merlin Sheldrake, for example, describes the fungal network or mycelium that interlaces with and meshes together the plant roots in soils in semiotic terms. The fungi send signals along their filaments, or hyphae, in the form of electrical pulses and can do so through the network over long distances. Because of such networked signalling, and because when fungal networks navigate through the soil they display directional memory, Sheldrake suggests that it is reasonable to consider the fungal mycelium as possessing intelligence.[6] Environmental journalist George Monbiot also describes plants and bacteria as communicating through chemical signals:

Microbes live throughout the soil, but in most corners, most of the time, they exist in limbo, waiting, in a state of suspended animation, for the messages that will wake them up. When a plant root pushes into a lump of soil and starts pumping out signalling chemicals and sugars, it triggers an explosion of activity. The bacteria responding to its call consume the rich soup the plant feeds them and proliferate at astonishing speed [. . .]. Plants speak in chemical languages that only the microbes to whom they wish to talk can understand [. . .]. The language changes from place to place and time to time, depending on what the plant needs. When plants are starved of certain nutrients, or the soil is too dry or too salty, they will call out to the bacteria that can overcome these constraints. Some biologists describe this as their 'cry for help'. In response to these chemical cries, a specific community of bacteria proliferates around their roots.[7]

Whilst notions of 'speaking' and 'crying' might be considered an anthropomorphic projection of ideas of human language, they do capture the more semiotic processes at work, in which some organisms are sharing signs to which others are responding. And for all Monbiot might stand accused of anthropomorphism, biologists and others who read such signs as cues and signals in a complex interplay of innate, pre-programmed behaviour might themselves be accused in turn of conceiving of 'meaning without meaning', drawing on mechanistic metaphors; of being 'mechanomorphic'. Are there alternatives?

In soils there are multimodal signals in play. Some examples highlight electrical and chemical signals, but signalling through touch and pressure also occurs, for instance as plant roots push through soil, or as insects, earthworms and other animals such as rabbits tunnel and burrow. Yet as in these examples, and as in most biosemiotics analysis, such signalling or signing is envisaged as prelinguistic. Signs such as chemical or electrical flows are treated as iconic or indexical signs, lacking a symbolic dimension. Signing processes are for the most part considered in isolation, without attention to how signs combine in grammars, or in wider packages of signification that build up over time— aspects that would be important in a structural-biosemiotic analysis of the communicative world of soils. A more structural-biosemiotic analysis could consider, for instance, how the touch and movement-based signing processes of animals—worms, insects and crustaceans, spiders and mites, centipedes and millipedes, molluscs and others—are picked up on both by each other, in the different layers and specific conditions they inhabit, and in adapting to significant juxtapositions. It could consider how signs (of ant chemical secretions, perhaps) combine with the chemical and physical signing through which animals digest and process vegetative matter. The combination might create an open soil structure replete with spaces and tunnels, into which rotting vegetative matter could perhaps move, signalling its presence to bacteria who respond by breaking it down further. If, as they do, such signing acquires sense in relation to juxtapositions and sequences, it can be considered linguistic. All these combinations might build up over time to develop a soil characterised by its thick, loose humus layer; itself an epiphenomenon, perhaps, but also a sign that might be apprehended visually or through touch (a dark soil, a soft soil) by other soil-using beings: a rabbit for example, or a farmer. The meanings conveyed by such a sign will surely depend on who the being is, and on the field of meanings particular to it. To a rabbit, it might signify a soil in which one can burrow (as opposed to others unsuitable, or others better still for a warren). As in this simple example, meanings are created and shared amongst soil beings through combinations of signs (in grammars or syntax) and wider packages of signification, the only question being the balance in these fields of meaning between their emergence through experience and their more biological 'anchoring'. This is a very different way of conceiving of what soil ecologists have understood in mechanistic ways as soil ecosystems; instead, we see a multispecies soil assemblage held together, and changing dynamically, through semiosis and the sharing of meanings. Soil communities become communicative communities.

Human–Soil Conversations and the Significance of Metasigns and Narratives

People are part of such communicative soil assemblages, not separate from them. When we humans interact with soils, whether intentionally—cultivating crops or digging for useful minerals or graves, moving or changing soils, as when building, or intervening to improve or care for them—or less intentionally, by walking over or dropping waste on them, we become part of living soil processes.[8] A structural-biosemiotic approach probes how this participation is communicative; how people become part of more-than-human conversations with soil that change both the soils and themselves. Cultivation is conversation; and so are myriad other kinds of human–soil relations.

In engaging in these conversations beyond the human, people are surely confronted with their inability to pick up sensorially on many of the multispecies signing processes going on amongst other soil beings. Most of the messages and signalling pathways involve senses that are inaccessible to humans without the specialist technologies: chemical or electrical signals, for instance. Many take place underground, out of immediate sight or beyond touch. Yet it is not necessary to apprehend the details of multispecies soil semiosis, or all the meanings that other soil beings might be sharing or picking up from them, to participate in soil conversations. People can instead pick up on higher-order epiphenomena and their significations: significations, somehow 'prepackaged', that result from the combinations of signing processes. These can be conveyed by, for example, the presence (as we illustrate below) of a particular soil-dwelling animal—a specific species of termite, perhaps—or a colour, or the textured feel of a soil, as well as other features such as a soil's smell. Such is the stuff of formal soil classifications, whether developed by soil scientists or by ethnopedologists. But as we will uncover and illustrate, they are better considered as emergent from combinations of signing processes in lively soil assemblages that build up over time and become significant within fields of meaning. They then operate as metasigns (always in relation to others) that convey the current quality, or significance, of a soil; or even convey a narrative telling of how that soil came to be of that quality. People 'read' such metasigns in relation to juxtapositions and narratives, and respond to them in ways shaped by their experiences: that is, in languages. Moreover, their responses subsequently further shape the soil and the meanings it conveys, pushing conversation forward. In this view, people pick up on and work with the ways that soil grammars are emplaced, spatially and temporally, in ways that are sometimes malleable (or indeed the result of past human–soil relations) and sometimes less so. Learning and working with soil language in this way is, in turn, important to living and working well with soils so that both people and soils flourish.

Metasigns in Soil Conversations

The presence of particular animals in soil can be picked up as a metasign. We authors can start to exemplify this in our own South Downs field where, as described in the last chapter, we have been trying to establish a small woodland. The soils here are shaped by their emplacement on a gentle midslope between the heights of the Downs with their exposed chalk geology, and the clays of the valley bottoms. In our field, beneath a thriving but thin topsoil, a layer of almost impermeable clay, about thirty centimetres deep, overlies chalk. This shallow clay layer became apparent to us as soon as we started to dig, as was its hardness and density—signs we picked up through feel. With some knowledge of the site's history, we were able to read this soil state as being in part the result of long geological processes, and perhaps partly of decades of compaction by grazing sheep and haymaking vehicles. We were concerned that the young tree roots would have trouble penetrating it, and also be disrupted by the rabbits that also inhabit and burrow in certain looser soils near ancient paths. We were encouraged, though, by the presence of an enormous number of earthworms: a sign, as we were assured by the long-experienced

Wilding Network volunteers who helped us with the planting as the Covid-19 lockdown loomed, that this soil would be 'good for trees'. The pink and grey worms that we revealed every time we split open the soil and inverted a clod to bury the roots of another seedling became, to us, a metasign conveying a meaning that gave us confidence that the trees would take root—as indeed most since have, excepting those dug up by rabbits, gnawed by voles and stunted by drought.

The presence of worms is a recurring metasign for gardeners and creators of compost. Their very presence, picked up on by people through their stand-out colour, shape and movement, is often read as a sign of a soil that will be rich and fertile. One does not need to know the details of the myriad intricate multispecies signing processes that the worms themselves are involved in, through their burrowing and digestion of vegetative matter, to appreciate the result. A rich soil is signalled by its multitude of worms. Our children's grandmother, who knew the soils she cultivated before the mechanical and chemical revolution in agriculture since the 1940s, recalls the soils predominating then as heaving with life. The development of a compost heap, or of such fertile patches, tends now to be restricted to the soils of gardens, where gardeners engage in a kind of conversation with soil and its worms: the gardener finds ways to encourage worms, by adding vegetative matter, and values their growing populations. As Maria Puig de la Bellacasa emphasises, this attentiveness and encouragement of worms has become part of the formalised practices of permaculture, as well: in permaculture training, gardeners are urged to appreciate worms as signs of the fertility they help create as soil particles and rotting vegetation pass through their guts to produce worm casts that nourish and structure earth, and encouraged to 'check if a pile of compost is healthy by attending to the population of pink sticky worms'. As Puig de la Bellacasa points out, this is in fact an everyday set of practices (but that we would now conceive of also as communicative practices) that are part of gardening and composting. She draws attention its emotional, affective dimension: 'becoming able of [sic] a caring obligation toward worms as our earthy companions in this messy and muddy way is nurtured by hands on dirt, curiosity and even love for the needs of an "other".'[9]

In the areas of Guinea where we have learned from farmers, people often read fertility not from the presence of worms, but from termites that play a similar role, as well as from the colour, texture, scent and history of the earth. The distinctive termite mounds are readily picked up on visually in the savannas, forests and fields once fallows have been cleared. Farmers we worked with often prefer the soils near mounds for planting particular rice varieties, as well as crops used for condiments such as local peppers or bitter aubergines, and select these soils over and above those in the rest of the field, where they sow other rice varieties. When dividing fields between each other, the dividing line is often through termite mounds, not just because they are visible, but because neighbours would otherwise compete over them. People draw on their experience that particular cultivars and varieties thrive in these soils, observing the crops in relation to these and the everyday hypotheses tested out in 'conversations' extending over years. Advice from parents and other farmers plays into this more-than-human conversation, entangling crops, termites, soils and friendly advisors. Farmers have no need to know or appreciate the multiple signing processes between termites (with their complex social hierarchies) and between termites and fungal networks, minerals and water that have led over years to the building of a mound, and to the particular soil and water characteristics in and around mounds, including the augmented clay content (that termites bring from subsoil), friable structure and raised humidity. If asked explicitly about the qualities of termite mound soils, farmers might explain also that water enters them well. But more significant is the presence of the mound itself, which they are able to read as a metasign conveying the meanings of fertility: a swelling of the earth echoing pregnancy. The mound conveys, too, a narrative that such crops have done well in such places in previous cultivation cycles, suggesting that they might do so again; a narrative that condenses and conveys meaning based on a package of past experience.

People in this part of the pluriverse often refer to termite mounds as termite 'villages'; as places of termite social life which they consider not just analogous to, but interconnected with, human social life.[10] This is part of a set of framings in which human and non-human processes relevant to fertility, reproduction and flourishing are less conceptually separated, than integrated within more unified socio-ecological reflection. Here it is conceivable that disruptions in human parts of the ordering—having sex in the wrong place, or with the wrong people, for instance—can have direct consequences for the fertility of other human and non-humans alike: crops might fail and the bush might 'tie' itself, disrupting usual animal movements and leading to hunting failure. Particular more-than-human conversations, such as those between people, soils, crops and termites in the course of cultivation, are part of this wider ordering, or what we might understand better as grammar and higher order narrative structures. In this context, happenings within termites' social world can affect people's, and vice versa, and this shapes the meanings of particular signs. People might read flourishing crops on a termite mound as a sign that all is well in the termite village, and in the human village world too, conveying a broader meaning of mutual flourishing. Within this grammatical ordering, the presence of termite mounds is also a sign of correct seasonal rainfall cycles, inasmuch as rainfall is partially controlled through the 'rainbow snake' or *ninkinanka* that is observed as a rainbow in the sky, but starts and finishes, people say, in a termite mound. Within this narrative ordering there is a logic to the association between termite soils, their moisture retention, and crop flourishing that goes beyond the details of what happens in soils, to extend to wider seasonal cycles and their more-than-human ordering, and in which termite mounds on the ground and rainbows in the sky can both be read as metasigns conveying that all is well.

Alongside soil animal life, colour provides another field of metasigns through which people pick up meanings about soil conditions or quality. Colour is a stock-in-trade of scientific soil classifications, whereby comparison with standard soil colour charts is used to infer the relative presence in a soil of white/grey mineral matter, and the main pigmenting agents of organic matter, iron, and, to a lesser extent, manganese.[11] Colour is highlighted too by scholars of ethnopedology, revealing ways that colour differentiations are important in local and Indigenous soil classifications. It provides an indicator from which soil qualities and the pedogenic processes that led to them are supposed to be inferred. Within our structural-biosemiotic framework, such classificatory approaches treat colour as a metasign that acquires meaning in a field of such signs: as furnishing indicators that convey the outcome of a whole set of signing processes, from those involved in mineral decomposition to the micro- and macrofaunal activity implicated in vegetational decay and organic matter creation. Each colour, in this way, conveys meaning through packaged signification but in relation to other colours (and qualities). Thus we need to go beyond static approaches to soil classification and categorisation, to explore how soil colour acquires meaning as part of a wider field of meaning, and how soil colour and its meanings are created—and indeed transformed—within human–soil conversations.

The significance of soil colour as part of more-than-human conversations is relevant in the everyday lives of many amateur gardeners. Many look with satisfaction on the dark soils of a vegetable or flower bed, or even a window box, to which they have recently added compost, for instance, and imagine their plants thriving happily there. Gardeners may read meanings such as richness and fertility from this dark colour in a field of meaning in which pale or red or chalky soils lack these things. The addition of compost becomes part of a conversation with soil and plants in which one adds the right amount at the right time of year, and attunes planting to suitable patches—aware that some plants will thrive in rich dark soil, and others elsewhere. Such attunement comes through experience, perhaps assisted by guidance from gardening books and websites, and advice from tutors, friends or the radio. When people say experience is the best teacher, they collapse the linguistic barrier that has been placed between two domains of conversation. Composting is a conversation, taking place over seasonal cycles and involving soil changes signified, in a packaged way, through their colour, and its contrasts with the colour of non-composted soils. Composting well, gardening well so that soils and plants flourish, depends on developing this soil language.

In the Upper Guinea forest region, soil transformations conveyed in colour change take place through more-than-human conversations that unfold not just through seasonal cycles, but over timescales of decades.[12] Many rural villagers, when asked, 'What kinds of soils are there here?', invariably reply that there are red and black soils, sometimes adding white soils as a third category. This broad colour categorisation is also identified in ethnopedological studies elsewhere in West Africa and is associated by some anthropologists with the primacy of red, black and white colours in Mande languages.[13] Yet for villagers here, soil colour, and especially the contrast between black soils that are found in particular places and red soils that are more ubiquitous, conveys meanings that refer to emplaced histories of human–soil relations and conversations. Black soils are most commonly found in rings around villages, and under existing and abandoned settlements. They form through the everyday depositing of waste from women's kitchens—food and ash and char from cooking fires—and the agricultural processing conducted in the village. As an elderly woman in the village of Wenwuta in Liberia recounted, for example,

> What makes the soil so rich? The dirt we put there and burn over and over for a very long time will change the soil. The black soil is rich around the town because the things we throw there: rice straw, fire ash, other materials. Soils on the farm are not as rich as those around the town as they do not have things thrown on it like in the town.[14]

Everyday waste throwing—which could itself be elaborated in more-than-human semiotic terms—can create thin layers of black soil within a year or two. Women often cultivate vegetable crops on these convenient behind-kitchen patches, and burying the waste from these moves the conversation along, increasing the darkness and depth of the black soil. Over a decade or so, they find, a deep layer builds up, forming a ring around the village. Farmers sometimes deliberately create patches of such soil in a similar way in parts of their fields conveniently close to a farm hut or regularly cultivated rice swamp, and hasten the process of transformation by adding extra char and burying waste by 'mounding' the soil. This soil transformation is durable, and black soils persist in these locations even after a village or farming encampment is moved or abandoned. Village territories in this region thus show a pattern of patches of black soil, associated with current or old settlement sites, with red background soil in between. The emplaced patches of black soil convey, to villagers, meanings that are both about soil quality and about the histories that created them; histories of cultivation, but also of the family and political relations that have shaped the use of particular farm sites and the settlement and moving of villages. That Kuranko villagers call black soils *tombondu*, literally 'land of an old village', and will refer to soils that they have recently transformed to black as 'having become like *tombondu*', highlights the importance of settlement histories in the meanings conveyed by soil colour. Black soils thus acquire packaged meaning (in contradistinction to other soils and their meanings), and do so in juxtaposition to other types in the landscape; such juxtapositions then convey a narrative—tell an emplaced story—both of soil transformation, and of people's settlement and political relations.

These histories and legacies of soil transformations in dark earth in West Africa are echoed in other parts of the pluriverse. In the Amazon, similar soils have long been known and studied by historical ecologists, archaeologists and soil scientists, who recognise them as *terra preta*. Patches underlie areas of Amazonian forest that were once assumed to be pristine: the legacy of soil transformations by Indigenous populations up to two thousand years ago, long prior to European colonisation.[15] Here, the pattern of 'anthropogenic dark earth' patches under what is now thick forest relate a narrative of dense settlement and peopling, followed by a subsequent apocalypse associated with the Hispanic conquest and the devastation wrought by the epidemics it unleashed.

Soil 'feel' is a third kind of metasign that can condense multiple aspects of soil quality and the processes leading to it. Like colour, this is a component of formal soil science classification systems. These classify a soil's texture according to its relative proportion of sand, silt and clay, with their different sizes of mineral particles, producing a series of soil types.[16] Because the size of mineral particles in a soil is not subject to ready change by cultural practices, soil scientists usually consider texture as a permanent, basic property of soil; compatible, then, with a mechanistic view of soil assemblages. Yet for many farmers and others who live with soils, texture, or feel, also communicates meanings that are about soils as something living, in which mineral particles and soil biological life, human and non-human, codevelop and change through semiotic processes: through conversation. Farmers, gardeners and miners alike are used to 'testing' a soil by feel, finding it perhaps hard or soft, crumbly or clumpy, sticky or powdery—or any other felt quality picked up by hand, spade or hoe. The feel might be the outcome of multiple processes in a lively soil assemblage, involving signing amongst macro- and microorganisms, plant matter, mineral elements and more; yet it is through the packaged result of these processes, resulting in the metasign of a particular feel, that they appreciate the soil's character, and possible purposes.

Kuranko farmers in Guinea would thus refer to soils 'becoming soft' through repeated mounding and burying of agricultural wastes, year on year. They would draw a contrast between the 'hard' soils of 'new' (unworked) land

and the 'soft' soils of 'old' (long-cultivated or settled) land; a structural contrast that also references the different histories of the respective sites. Soil texture, or feel, can thus relate a narrative; tell a story. These farmers would also refer to the 'oiliness' of a soil; an aspect of feel that, again, was a sign of repeated human use and everyday life in cultivation and settlement. The dark soils of old village sites were often distinguished as 'oily', and thus highly fertile, a characterisation that aligns also with wider framings in the region that link oiliness and fertility more generally, including in people—articulated for instance in the oiled-up appearance of women initiates at the 'coming out' ceremonies celebrating their newly fertile adult status, and in the shiny (oily) black masks that are part of initiation society activity, securing the fertility and reproduction of the social order.[17] Oiliness in these human social activities is a sign of fertility, as oiliness in soils is a sign of their fertility; a metasign that speaks also of the interconnectedness of human and soil fertility in this part of the pluriverse.

Metasigns can convey meaning, or relate narratives, in their combination. Colour and feel, for instance, might each bolster the other's significance when aligned. Thus for Guinean villagers, a soil that was both black and oily was the clearest sign of an area of *tombondu*. For *terra preta* soils in the Amazon, the layered human–soil histories involved in their making are conveyed in both their dark colour and their open texture. Colour–feel combinations can also signify narratives of soil decline, and the processes precipitating them. Thus the pale powderiness of soils found in some areas of the UK can be read as narrating a story of decades of mechanised ploughing and chemical application, shaped by the political and economic imperatives of industrialised production.

There is a communicative dimension also to the objects found in soil assemblages. Human-made objects in soils might initially strike one as matter out of place, standing out for their disruption to the grammar, and to what one would expect. Such objects might carry immediate meanings: the plastic bag in one's compost heap or village dark earth that speaks of disorganised household waste management, or of plastic pollution, for instance. Yet they can provide further signs that convey narratives of human–soil conversations and the social lives that informed them, often over long periods. When we found pieces of grey clay pipe in the soils of our garden, for example, we explored further and learned of the pipe factory that once occupied a neighbourhood site nearby. In old settlement soils in Guinea, relics of those who lived there before are sometimes found—the shapes of houses, pieces of pottery, even ancestral graves. In Kissi-speaking regions, villagers sometimes dig up small stone figurines in fields, which they call *pomdo*; they read these as a sign of earlier habitation and also of the fertility of the site—as the former is in itself inevitably productive of the latter. Interpreting the meanings of artefacts in soils is, of course, the stuff of the whole discipline of archaeology; yet a structural-biosemiotic framework brings both rapprochement with and a new dimension to archaeological studies, by considering their objects and methods in terms of the language of soil assemblages.

Emplaced Soil Conversations

In probing the language of soil assemblages in this way, it becomes clear how emplaced it is. Grammars and signs acquire and convey meaning in relation to place; places shaped not just by what soil scientists often consider as background geological and topographical conditions, but, crucially, through the conversations and histories that have unfolded there. The result is often that landscapes contain patches of different kinds of soil. An appreciation of soils as communicative assemblages reconceives soil cover as an emergent mosaic: a 'patch dynamics' of interest to social scientists and ecologists alike, but whose communicative dimensions have thus far been underplayed.[18] There is also a temporality to the grammars and signs in more-than-human soil assemblages that emerge through conversations unfolding over a diversity of timescales, and in their combination. Thus conversations in the course of a seasonal cycle may shape those taking place from year to year, and from decade to decade; and communications that happened decades or even centuries ago (regarding how ancestors settled a site and transformed its soils, for instance) leave signs that beings today might interpret and build into present and future conversations. Maria Puig de la Bellacasa recognises these multiple overlapping timescales in more-than-human soil relations, and the ways they sit at odds with the temporalities of industrialised soil management. As she puts it, soil care (as practised in many cultures and places) involves multiple temporal rhythms, yet these, she observes, are often subordinated to the 'linear temporalities of technoscientific productionism'.[19] These linear temporalities, she suggests, encompass a fundamental contradiction between soil seen as, on the one hand, an entity slowly renewing through a combination of the long, gradual geological processes that break down rock and the relatively shorter ecological cycles by which organisms and plants, as well as humans growing food, decompose materials that contribute to renew the topsoil, and soil seen, on the other hand, in terms of accelerated, hyper-short timescales focused on intensified productivity through technological solutions. Our perspective complements this, by showing how multiple temporalities emerge through communication, and how temporality itself comes to be part of emplaced soil juxtapositions and the syntax or grammars they follow.

Human–Soil Communication and Mutual Flourishing

Considering soils as living assemblages in this way, and probing the communication beyond the human through which human and non-human beings interact, helps elucidate ways that people can work with living soil processes in ways that enable all to thrive. Appreciating the language of soil assemblages enables conversations that shape soil processes in ways that are productive both for people, for purposes such as cultivating crops, and for other soil beings (enabling soil animals, fungi and bacteria to thrive, for instance), in ways that feed back over time to create a beneficial spiral. Such positive interactions between humans and soils, working with rather than against complex soil ecologies, can be found in a vast array and diversity of Indigenous and local practices in living with and using soils across the world. Some are well established and elaborated as systems, such as those of regenerative agriculture or permaculture, while some are informal and everyday: what goes on in many backyard gardens and fields around the world. At least as far as formally institutionalised principles are concerned, however, communicative dimensions are barely acknowledged. For instance, of the thirteen principles of agroecology, endorsed by organisations from the United Nations Food and Agriculture Organization to national and transnational peasant organisations,[20] none refer explicitly to communication. Principle 6 calls for 'Synergy'—that is, to '[e]nhance positive ecological interaction, synergy, integration, and complementarity amongst the elements of agroecosystems (plants, animals, trees, soil, water)'. Yet the conceptualisation is mechanistic, of interactions amongst elements of a system, not of communication amongst living beings; not of communication that admits to and helps explain the affective relations that people have with place—with nurturing lives in relation to a place.

There is, however, a more communicative emphasis in permaculture, a movement originally associated with founders Bill Mollison and David Holmgren in the 1970s, which aims to enshrine the triple motto of 'care of earth, care of people, return of the surplus' through '[c]onsciously designed landscapes which mimic the patterns and relationships found in nature, while yielding an abundance of food, fibre and energy for provision of local needs'.[21] The principles and guidance that circulate in permaculture networks and training courses advocate attentive, multisensory engagement between human practitioners and soils, as in this example from the Green Connect urban farming project:

1. Observe and interact: Before you take action, start by taking some time to observe what's happening. If you want to build a garden, watch the space and see which parts get sun and rain and which parts get the wind or shade. Your job becomes so much easier when you can work with nature,

rather than spending time and effort tending to plants that are growing in the wrong spot. [...]

4. Apply self-regulation and accept feedback: Your garden is its own ecosystem, and some of your interventions might have negative effects on parts of it. Watch your garden closely and listen to what it is telling you. Once you've heard the message, accept it and make changes.[22]

Mollison, one of the founders, speaks of embodied immersion in soils through 'TAPO—thoughtful and protracted observation' before acting on the land and its processes. Yet listening and observing are conceptualised as being done by people, to and of soil processes—as if the two were not entwined. Puig de la Bellacasa, by contrast, considers permaculture, amongst other approaches she characterises as 'matters of care', as a set of more-than-human practices. She notes that some renowned practitioners emphasise that it is a matter not just of acting on or mimicking the environment, but of humans actually 'becoming nature working'. She interprets the 'culture' in the term permaculture as indicating also the cultivating of communal practices over time within a community of human and non-human beings. Yet even in this work, we think, communication beyond the human is underplayed. Seen through a structural-biosemiotic lens, as we have emphasised above, soils are themselves living, communicative assemblages in which all beings, human and non-human, are engaged in the creation and sharing of meaning. Both everyday human–soil interactions and those that stem from formalised approaches such as permaculture or agroecology can be helpfully reconceived in this way.

Appreciating the communicative dimensions of more-than-human soil relations also highlights how these can be affective, pulling on people's emotions. The smell of a soil might bring powerful memories; the feel of a soil might bring pleasure, or comfort, or simply the distraction from worries that comes from 'getting earthy'. These are not sensations of meaning that are merely iconic or indexical, but quintessentially acquire their significance as metasigns, making sense in fields of meaning and juxtapositions that are laid down in and emerge through experience. Many people speak of gardening as therapeutic, and it is part of the repertoire of 'green prescribing' offered for the enhanced well-being of stressed-out modern urbanites. As a participant described of a permaculture training course,

> being in the fields and woods, doing this work with the soil, the water, the plants, getting muddy, learning about more than human working patterns and how to foster abundance [....:] the affective feel that has remained with me for a long time after this training was a sense of renewal of collective hope and joy in the face of a frightening and often depressing world.[23]

This affective power of immersion in—conversing with—soil can be understood, at least in part, as arising because it involves immersion in a living web: a communicative web, in which one's self is decentred and subordinated to wider, more-than-human relations.

Soils, Land and Entanglements of Identity

Affective dimensions of human–soil relations are also significant insofar as they become entangled in wider human social life, ethics and politics. For instance, agroecological and permaculture practitioners and analysts, as well as farmers, gardeners and the researchers who study them, describe how an ethics of care—for people and the earth—can emerge from people's relational encounters with the living soils that sustain them.[24] Anne Therese O'Brien thus describes how for Australian farmers involved in pasture-cropping, 'new dimensions of soil flourishing become evident, and the distress of soil ecosystems is rendered ethically acknowledgeable'.[25] Indigenous peoples and their advocacy movements claim identification with and attachments to land—attachments that go beyond tenure and rights, to encompass histories of mutual living with soil beings and the sense of community and affection so forged—and which can become part of struggles for justice.[26] Such attachments are not simply with human grave sites. Discourses about (Indigenous) people and land often portray things in terms of 'sacred ties to the land' linked to a particular human 'culture'. Yet we might reflect that these links are quite as intensely communicative as human social links. Lands share meaning in innumerable ways, and these are picked up on by people (and other beings) in ways that may create more-than-human communities.

Mutual identification between societies and soils can of course divide and exclude. A dramatic instance is in Germany, where certain strands of environmentalist thought became associated with the idea that there is a deep (some characterise it as 'mystical') connection between people of German blood and the soil of the lands they inhabit. As Janet Biehl and Peter Staudenmeier describe,[27] this 'blood and soil' complex can be traced back to nineteenth-century thinker Ernst Moritz Arndt, whose foundational ecological thinking was tied up with virulent nationalism and xenophobia, couched always in terms of the mutual well-being of the German soil and the German people. These ideas were picked up and developed by movements which 'preached a return to the land, to the simplicity and wholeness of a life attuned to nature's purity',[28] and then by the German youth movement the Wandervogel (wandering free spirits) whose members turned to Nazism in their thousands in the 1930s. National Socialism appealed to many young German nature lovers, as strains of ecological thought were woven into Nazism and into the personal habits and writings of Nazi leaders, including Hitler and Himmler. Associations between organic agriculture and xenophobic currents emerged in the UK as well, at the heart of the organic agriculture movement, and persist even to the present.[29] What tends to be underplayed in analysis of these interconnections are the ways the power of such politics is rooted in meanings that encompass a conception of (certain) people as part of the living soil, and the affect this produces.

Land and soil can also unite people. For example, anthropologist Philip Winn describes how inhabitants of the Banda islands in Indonesia during the twentieth century disclaimed any suggestion that they were Indigenous or had ancestral connections with the land (and indeed, they all had origins elsewhere), yet for these new mixed populations, Muslims and Christians alike, as cited by Amitav Ghosh,

> the islands were a 'sacred territory' (*tanah berkat*) anchored by the spirits of founder figures or *datu-datu*, who were thought to have arisen from the land. Even after the islands' original population had been mostly eliminated, they continued to animate the land, living on as 'emplaced spirit forms, maintaining a watchful presence and exerting a powerful influence over their original territory'.[30]

Ghosh thus suggests that it is the very vitality of land that can draw people together, co-creating more-than-human communities. The violence that overtook the Banda islands in the twenty-first century, during which most Christians fled, was therefore also a rupture in these relations that tied people to animate land.

Likewise, working together with living soils can co-create human social communities and solidarities. When people come together on allotments, such as one finds in many British cities, there is a creation of human community through numerous communicative acts between people: those involved in sharing knowhow, help, tools or produce; in chatting about and comparing cultivation experiences; in negotiating tenure and boundaries. This might bring together people of very different backgrounds, ages, other interests. In the United States, coming together in gardening projects has enabled the forging of solidarities and creation of human communities that are significant as refuges from, and in struggles against, racial injustice.[31] Yet we might reflect that these are more-than-human communities too, and the sense of unity that comes from shared cultivation in a place also comes from the shared experience of participating in conversations with the soil and its myriad animate and inanimate beings, participating in and thus co-creating a more than-human community.

Soils, as we have illustrated, can thus be envisaged as more-than-human assemblages, constituted through all kinds of communicative processes between the human and non-human beings that are part of them. Yet as such framings of soils as lively assemblages, and recognition of the importance of this to sustainable human–soil interactions gain ground, so the tensions grow in relation to framings that deny all this. These include those dominant in much industrial agriculture and its earlier, underpinning soil science, where a conceptualisation of soils as consisting of dead minerals still predominates. They also include those that associate soil quality and fertility with an undisturbed 'nature' that human, plant and animal interactions only disrupt or undermine. Both these positions can support and justify chemical fertilisers as the key route to maintaining and enhancing soil quality and fertility. They can also justify approaches to soil management and conservation that remove human interactions altogether: leaving soils to 'rewild' under so-called natural vegetation whilst food for humans is produced elsewhere—in labs, greenhouses or focused, technologically enhanced 'smart' plots;[32] or managing them to sequester carbon at maximum efficiency and scale.[33] These approaches can become part of a wider set of human–nature separations, to which we shall return in our final chapter.

10
Seas

Seas are assemblages, but what dominates is less their life forms than their elements: water, salt, wind, and the sand and stones that their waves beat against. Underwater life is more elusive, or at least harder, for people, to grasp. This makes human immersion in and communicative interaction with seas a challenge to describe, and the dilemmas we explored in relation to forests and soils, where there is apparently no mind, yet a communicative encounter takes place, are even more pronounced. With sea too, however, codeveloped fields of meaning, grammatical structures and their packaging are at work, and with lively and vital inanimate elements that emerge as one one moves beyond the presuppositions of Enlightenment science.[1]

Our ethnographic examples in this chapter begin with cold water swimming in south-east England and expand to the global practice of surfing, before turning to the worlds of fishing communities in coastal Brazil and East Africa, and to underwater communicative life off the coasts of southern Africa. How structural principles manifest depends on both the particular physical emplacement of different seas, and what beings are relevant, across a pluriverse of worlds in which winds are sometimes personified and sea beings may include serpent-spirits. Our exploration of seas in this chapter foregrounds some particular angles in structural-biosemiotic analysis, including the ways that technologies become imbricated in communicative assemblages, and the significance of higher-level narratives—here building further on our consideration of forest and soil narratives. More-than-human sea assemblages are thus revealed as being held together by webs of structured semiosis, contrasting with a mechanomorphic portrayal of marine ecosystems and politico-economic debates that reduce sea creatures to resources and commodities, and emplaced habitats to fisheries.

Elemental Communications and Sea Immersions

Cold water swimming all year round has recently become increasingly popular off England's south coast, including in our home city of Brighton and Hove. What is it that impels people to enter that cold water, even in the depths of winter? Participants assemble on the beach in the morning and meet up with fellow swimmers, friends and cold water acquaintances in what has become a human social activity; but when entering the water, each knows they will enter a world so bracing that it will take them into the moment and away from human-social worries. Such immersion, too, induces an element of fear, and a reframing of the narrative configuring the signs that our senses pick up from the sea; a genre involving the vulnerability of becoming a small bit-part player amidst the vital forces of nature. But these swimmers are among friends, sharing a common experience, each switching narrative in parallel with human companions. Afterwards, they talk of the feeling of the sea being in control, and oneself not, as the body is pushed and pulled this way and that by its forces, tugging on arms and legs and flexing the back, a lively force that plays on our proprioception. Swimmers sometimes describe a sense of dissolving into a wider elemental world, and a world in motion and of emotion that dispenses with the vanity of humanity.

The desire for such immersion that cold water swimmers express is also a desire for communicative connectivity with the sea. Yet that communication is not simply embodied, but happens in a structured way as one plunges in. The cold sea feels entirely different from what one would feel on dry land, and the transformation is instantaneous: the tingling cold ripping through the body and contrasting with all the signs it shares meaning with—not just warmth, but extending to introception of heart rate and weightlessness, proprioception of movement and skin sensations of wetness, the darkness of depth contrasted with the reflected light of the sky, the viscosity of water as compared with thin air in which one moves easily, the movement of water as shaped by the wind and the waves, and the sharp smell and sting of salt in face and eyes. The cold water is not a sign alone, but a sign that acquires sense in a field of juxtaposed meanings and experiences. It does so in ways highly attentive to syntax, so that on plunging in we immediately discern if something is awry, out of place, but are prepared no less for the switch of narrative it brings on us, as we hanker for it. This is a potentially dangerous immersion, and life depends on reading the sea correctly. The swimmer experiences a plethora of signs, but reads and interprets them in packaged ways rooted in experience, in acquiring familiarity with the sea; a conversational familiarity.

The sea is not one thing alone, but emplaced. The shape of the coastline, the make-up of the shoreline in terms of rocks or buildings, stones or sand, or nearness to the shore as opposed to being out on the high seas—all shape the assemblage. There are emplaced patches in the high seas, even, that can be distinguished by the nuances of their grammar: currents and the patterning of the waves, or doldrums where waves cease and all becomes calm. The sea changes temporally too, daily, seasonally and in irregular ways, being sometimes clear and sometimes not; sometimes calm and perfectly innocuous, and sometimes so rough as to feel dangerous to humans. The sea thus reflects the wider forces that shape it: of tides and the moon, of winds and weather, that manifest in any particular location. The state of the sea in which we immerse ourselves thus tells us something of these wider forces too.

Across the pluriverse, these wider forces might correspond to the weather fronts of modern meteorology, but they might manifest as the winds of the gods and spirits: perhaps, as among the Haudenosonee of upstate New York, as the mighty panther spirit of the west wind, the gentle fawn spirit of the southerly, the moose spirit of the easterly, or the fearful bear spirit of the north. They might be the water serpent beings that, as described by anthropologist Veronica Strang, have 'embodied ideas about non-human powers and water's elemental capacities to shape human lives' in a vast variety of settings. From widespread early histories in which 'serpent gods rose up out of primal seas to create worlds',[2] their more recent forms extend from African python figures to the plumed serpent Quetzalcoatl in Central America, to Japan's cloud dragons that variously influence and control the fluidity of water and winds, to the Māori story that conceives of the rise and fall of the tides as the breathing of the great sea deity Parata. Water deities, and the meanings they embody, remain powerful in many societies despite the dominance of scientific meteorology, hydrology and engineering.

The forces that move the elements can thus be framed as animate in varying degrees, as physical or spiritual, or combinations of these. Whether personalised or impersonal, they come to be characterised in relation to fields of meaning: in terms of signs and syntax, metasigns and the narrative genre of metasyntax, all acquiring sense in relation to each other and in conversational experiences with people and the elements. In such ways the grammar of any area of sea is connected into these wider forces within the pluriverse or cosmopolitan neighbourhood. Framings shape how we read the sea we are in, and the agency we attribute to any given element. Communication with the sea is with something lively, but how that liveliness is figured depends on who and where one is. Narrative framings differ even within our own home city of Brighton, and we can switch between them. Thus looking at the sea as one reads the weather on a phone app invites a meteorological framing of its waviness,[3] supported by the waves' visual signs and assigning a numerical windspeed. Yet once in the water some swimmers, at least, experience and describe what they feel in far more animate, agentive ways. Those immersing themselves in the sea are seeking some kind of connection, extending some kind of conversation, and this recaptures some of the liveliness that the meteorologists have bleached from the sea, allowing a more personal language to codevelop with the elements around.

Surfing Conversations

The surfing that Hawaiians and the inhabitants of the Marquesas Islands developed centuries ago amazed and shocked early American and European observers, but by now has become a sport, indeed a lifestyle, around the world's shorelines, through which many encounter the sea. Surfers describe engaging with the higher forces and learning the waves and currents, and they develop subliminal skills in spotting the signs of a wave and embodied skills in manoeuvring a board. This involves, however, learning a language in its widest sense: learning the orderliness of the waves, their juxtaposed varieties and their grammars. These are grammars with rhythms, over different timescales: of the day, the weather, the tide. Every wave is a little different, and there are terms for them; each has different implications, different meanings.[4]

In the littoral zone, the land–sea interface, many surfers are intensely aware of and moved by aquatic conversation, which is integral to their practice and experience. They come to develop intimate relations with the surf 'break'— the unique combination of saline liquid, atmospheric process and seabed topography that gives shape, value and a sense of power to breaking waves— discerned through repeated experience and practice. Through this, as anthropologist David Whyte exemplifies citing Irish surfers, a surfer comes to feel they belong in or with the ocean, becoming the sea's citizens: a 'saltwater citizenship'.[5] As the surfers described it to him, their bodily engagement, or communication, is with what they consider 'the original energy' of the sea, an energy which they refer to simply as 'the Ocean'. They experience this power as 'sometimes amenable, sometimes elusive, often ferocious and always unpredictable'.[6] Surfing integrates 'the dream' (mind) and 'the glide' (body) in something akin to coenaesthetics (the development of sensibilities through the senses), and that captures the semiotic dimension that we foreground in this book.

Drawing on interviews with many surfers around the world, anthropologist Jon Anderson captures how through surfing communication there is a sublimation of the self into a higher order, much as the horse-rider and mount become one momentarily, as we explored in chapter 4 above. 'The relational sensibility experienced through surfing', he writes,

> engenders a feeling of [...] transcendence. [...] Relational sensibility is the emotion felt within a human being, but produced through the co-

constitution of that human with, in this case, wave, board, fetch, geology, weather system, wax, wetsuit, etc. [...It is] the product of the coming together of a range of components into an assembled or converged state.[7]

In this 'becoming together', there is no longer a 'surfer' and a 'wave', but a 'surfed wave'. It is a dance with the ocean; intra-acting with and reacting to a constantly moving surface. As one surfer described to Leslie Kerby,

> I paddled fast to my left, angled toward the next wave, stroked and stood, felt the board accelerate, and pumped once and into my bottom turn—and then the world vanished. There was no self, no other. For an instant, I didn't know where I ended and the wave began.[8]

Surfers themselves variously express the emotions this convergence engenders in them: from happiness with 'smiles of unbridled joy', to a sense of awe in the face of powers that can be dangerous, or 'a sense of being a part of something that is timeless and much, much bigger than you. Waves have been breaking since there has been water on the planet and that knowledge can ground you in a period of unease.'[9]

Many analysts of surfing, and surfers themselves, describe these intense experiences in terms of a kind of spirituality or religion—and certainly as closer to this than to the hedonism that many associate with the surfing lifestyle. The activity is thus associated with a transformation of consciousness that comes about through something 'ineffable, beyond words': a kind of spirituality.[10] Something much the same is experienced by tightrope-walkers, climbers, and indeed, potentially, the rest of us who might not push ourselves to such limits. The physical, psychological and spiritual benefits of surfing have been deemed so profound that some have suggested that it 'should be understood as a new religious movement', coining new terms for this: 'Aquatic Nature Religion', or 'Surfism'.[11] Other analyst-surfers have argued that the intensity of surfing makes one mystical, likening the 'stoke', or high, of surfing to the concept of Nirvana or enlightenment in Buddhism to capture the profundity of a conversation beyond the human.

If there is a structural quality to the grammar of the sea, then there is also the potential for this to be disrupted by matter out of place, and a consequent grappling towards an alternative narrative genre (metasyntax) in which new order is to be found. For surfers, such matter can be literally 'matter' that should not be there: the human effluent, sanitary products and toilet paper that pour from pipes in many places along the UK's coastline, against which 'Surfers Against Sewage' campaign.[12] On a larger scale, media and campaign organisations express outrage at the 'fatbergs'[13]—boulder-sized lumps of congealed grease and gelatinous oil with toxic bacteria and often with nappies, wet wipes, condoms or even syringes attached—that escape sewers and wash up on beaches. Other fatbergs form in the ocean, where ships discharge sewage and oils such as palm oil. They are out of place in seas, on account of their look, their smell, their toxicity. Plastic bags, and what amount to enormous floating landmasses of ocean plastic, can also aptly be considered egregious specimens of matter out of place.

Emplaced Sea Narratives

Narrative genre can also be disrupted by events that overturn the normal order of things. A prosaic example occurred one evening on an English south-coast seashore where we found ourselves at the end of an intensely hot day when families had been swimming, paddling and paddleboarding close to the beach in a calm, gentle sea. The narrative had been one of peaceful seaside pleasure and sociality. But suddenly, without warning, the clouds blackened and a violent onshore wind got up. The calm sea turned ferocious in what seemed a mere instant. Swimmers and paddlers raced for shore, and within minutes the beach had entirely emptied, in a dramatic shift of scene. The sea was suddenly not a welcoming place. There was danger and fear: the conversation and its framing had changed; the order had shifted, and the unfolding narrative became one of hasty escape. Yet on reflection, such weather changes are themselves part of a wider order, easy to interpret, even predict, through meteorology and the physics of land–sea heat exchanges of the kind one learns about in school geography lessons. This could be considered a (very) mild version of a more dramatic event, whether storm, cyclone, or earthquake-related tsunami. Over a longer timescale, those used to living with the sea experience extreme events and their effects as part of the sea; within its language. To be familiar with more-than-human sea worlds is to be more rarely surprised by their liveliness.

Places in the sea have their own genealogies, which are in part histories of the events that have taken place there. These may leave material legacies: shipwrecks on the seabed, objects discarded from boats or shoreline pipes floating by. In a way similar to forests or soils, seas offer palimpsests, or living archives, of what has happened there before. This provides an alternative syntax in which signs such as sewage, fatbergs and plastic bags acquire meaning not as matter out of place, but as signs of the embedded genealogy of seas in our late industrial times. What is clear is that the sea offers these signs, in the form of objects and shadows glimpsed underwater; whether and how they make meaning depends on conversational encounters with the sea and the metasyntax—the narrative genre—invoked, be it more distantiated and technical, explanatory and historical, or a story of the out-of-place during immersion. It also depends on how such histories are kept alive in stories and social memories. For some people in coastal areas of Sri Lanka or southern India, the seascape will forever be entwined with memories of the devastating tsunami of 2004; for those in West New Britain, with stories of the volcanic explosion and collapse of Ritter Island and the mega-tsunami it generated in 1888 that obliterated their coastal civilisation.

Experience and communication with the sea as a place is mediated by technological possibilities. As surfing accounts emphasise, an assemblage is at work that comprises the surfer and the sea's elements and forces, but also the board, wetsuit and wax.[14] A swimmer's communication with the moving waves and currents can be mediated by flippers. And then there are boats, of many different types, and the array of technologies from navigation aids to radar, lights and lighthouses, buoys and radios, and technologies that 'see' the seabed in different ways and that detect and can shape direction, speed, wind movement, depth, proximity of shore. While the grammar of the sea remains, the language that humans speak as they engage with this grammar is not just human: it is human-plus-technological. Put another way, we and our technologies become part of the communicative assemblage. And this adds to the dynamism, as communicative patterns and possibilities change not just with the elements, but with the technology. A new, more streamlined boat; an engine replacing sails; a digital navigation aid: these change the way humans engage with sea assemblages.[15]

Communication Amidst Underwater Life

The conversations that we have been exploring so far are with the sea's elements, animate but perhaps not 'alive', although some across the pluriverse understand water and winds as living entities. Yet the sea englobes thriving ecologies in which aquatic organisms have sophisticated ways of communicating with each other, and that specialist scientists participating in the techno-biological revolution can learn to engage with. Octopuses are a focus for investigating an intelligence distributed through tentacles. Whales and dolphins are a focus for research on intelligence too: their sophisticated sonics shape and forge their hunting and social and cultural lives.[16] Finding signs with symbolic dimensions leads some biologists to qualify cetacean communication as language equivalent to that of humans, and whales and dolphins rank high amongst those animals identified as 'speaking', albeit through their particular sensory repertoire and in relation to their particular and oceanic *umwelt*. Biologists are showing how, assisted by technology and Artificial Intelligence applications, humans might learn to 'speak dolphin or whale'; to hack into and understand their signing processes and syntax in remarkable detail.[17] This rapidly advancing field is exciting, offering significant contributions in challenging the notion that only humans have language, and in envisaging human–whale interspecies communication. The exceptionalism that characterises much of this research, however, embodies a rule to which we in turn take exception, as it instantiates an intelligent-species-focused conception of language: a phenomenon that we see rather as a manifestation of ubiquitous structural-biosemiotic encounter. Yet with that caveat, such research does offer insight into how meanings (and sharing of meaning between humans and whales and dolphins) depend also on the ways multimodal signs are combined in grammars, and on how these are emplaced in space and time.

Take the dolphins inhabiting the waters off the coast of south-eastern Brazil, which regularly interact with people who fish for their livelihoods by bringing fish to the shallows and grouping them where they can be more easily caught.[18] The fisherpeople communicate with calls to the dolphins to an extent, but mainly through gestures, in the ways they move their bodies and nets and perhaps by lobbing a fish to say 'thank you'. The dolphins respond to both humans and fish; with a flick of the tail, a breach of the water, directional movement to push the fish closer together. A much wider field of signs is in play, beyond sound and with no need for detailed interpretation, in a conversation between humans, dolphins and fish and the wider assemblage, including waves and winds, that acquires mutual meaning in the moment. Signs are combined in a grammar that also involves place—the nearshoreline and shallows, accessible to people, fish and dolphins—and temporal rhythms, with this fish-gathering happening in the evenings, and at certain times of year. But while meaning is shared in the particular context of each evening's fish-gathering, it also draws on accreted experiences and the packaged meaning so generated. For the partnership between dolphins and fisherpeople and other elements in the sea assemblage here is long-term, and the signs, grammar and practices have been repeated over decades, and passed down through generations of dolphins and fisherpeople alike, co-creating what is a particular interspecies communicative community, or culture.

Dolphins and whales may be the poster-children for language amongst sea inhabitants, but people have developed enduring conversations with many others, and a structural-biosemiotic framework, by lifting the focus from the association between language and intelligence, helps to reveal this. Multisensory linguistic communication beyond the sonic occurs in myriad ways in the sea's communicative assemblages. Fish that move in shoals communicate by combining the visual with bodily reactions to the play of water around them as shaped by the movement of other fish: a veritable package of signs that keeps shoals together and enables their corporate agility in the water. Writing of cup corals off the coast of California, Eva Hayward coins the term 'fingereyes' to highlight how being and sensing, vision and touch, simultaneously work together in the sensorial world, the *umwelt*, of these aquatic creatures.[19] Whether in these or myriad further examples, a far wider field of signs is in play, as creatures communicate with each other and with the elements around in these aquatic conversations.

People participate in these conversations too. Charles Zerner explores how the calls across water of fishermen in Sulawesi enter a soundscape of the sea in which multiple creatures are already conversing.[20] Or take for example the so-called Great African Seaforest that lies off the western coast of South Africa, where the kelp forests and algal gardens are home to thousands of species of marine fishes, cephalopods, crustaceans and other shellfish, sharks and rays, corals, plankton and more. A group of scientists, journalists and film-makers involved with the Sea Change Project there have explored the communicative networks amongst them through methods of 'underwater tracking', revealing how their interactions involve visual, chemical, and touch-based signs, and the significance of these signing processes in constituting an undersea world of great complexity.[21] The Sea Change scientists' accounts rarely reference semiosis explicitly, yet this is what they are describing. Nor do they explicitly acknowledge the emplaced grammars in play, yet these are implicit in the patterns described, as creatures move, feed, breed and play in and between the layers of a triple-storey underwater habitat. In entering and exploring this world, through diving in the cold waters there and swimming with and amongst the creatures, Sea Change members have learned to read many of these signs, a process one described as delving into 'the biological mind of the forest' under water. They have become part of these semiotic webs, whether in one-off or repeated dives in the frigid waters. The much discussed 2020 film *My Octopus Teacher* documented some of these interconnections,[22] but its focus on the emerging companionship between the octopus of the title and the film-maker, Craig Foster, and their largely touch-based communication, gave less attention to the multimodal signing amongst other sea creatures, and how his swimming body became part of such semiotic webs: for instance, as it disturbed and re-routed fish shoals, altered the pattern of light and shade to which creatures were reacting, and more. As an avowedly peaceful explorer rather than a disruptive presence, Foster inevitably became part of the ongoing conversation amongst sea life, aware of (and worried about) intervention in the life and death of the octopus herself. Whilst initially his presence might have been picked up as matter out of place by the octopus, over time and daily immersions he became part of its normality, habituated within this emplaced assemblage's grammar.

The Sea Change project is explicitly about communication at a higher level, in that it is about storytelling; as its website claims, 'We tell stories that connect people to the wild, motivating them to become part of the regeneration of our planet,' and professional storytellers are amongst the project's human staff and supporters. Narratives abound in the project's written, visual and digital materials, from the recounting in film of the life and death of the 'Octopus Teacher', to narratives about sea creatures' lives and how people's encounters with them transformed their health. For example, Foster's regular immersion in the cold water of the kelp forest and its communicative life and his relationship with the octopus is told as a story of emotional and mental transformation and healing. One volunteer whose involvement in Sea Change coincided with attempts to recover from a painful knee condition describes how '[w]onders kept opening up. A whole garden of soft coral sea pens living in the sand. See-through chokka squid paralarvae coming into the shallows from the deep. Every day there was something new to focus on, distracting me from physical pain.'[23] Meaning emerges from metasigns and metasyntax—narratives that codevelop in relation to the coral lives and squid lives enountered, and with emotional, affective implications. Such narratives are human, but codevelop no less through conversation with non-human others. While the project emphasises (and publishes) narratives related by humans about sea life, the assembled sea creatures can also be understood as telling their own stories; recounting through their behaviour the outcomes of the accumulated experiences that have become their patterns of life. Whether humans pick up on these depends on their experiences and ability to read the language involved. For instance, the Sea Change volunteer quoted above may have read the emergence of squid paralarvae from the deep in terms of its novelty to him, but for the larvae themselves, or other creatures or indeed humans more accustomed to them, this movement from deep to shallows is presumably part of these sea creatures' own established narrative, with quite different meanings within *umwelten* unknowable to us, whether relating to feeding patterns, rhythmic migration, experiences in interaction with other creatures, or more. Picking up on the stories of the sea assemblage is better done by learning not just its signs, but the emplaced grammars these are part of, and the ways these have built up over time.

There are limits to undersea more-than-human communication. While some humans immerse themselves in this animate communicative world, many of us hardly know it. It is largely through documentary television, increasingly enabled by underwater camera technology, that it is revealed, but then in ways structured by the particular narratives that film-makers impose; the stories they want their films to tell.

We are often tempted to consider the limitations to communication caused by misalignment of senses between humans and other beings, and this becomes a particular challenge with sea beings, whose underwater *umwelten* are so different from those on land and in the air. But senses can be attuned and adapted, and technology can become part of the assemblage, facilitating alignment. An interesting example comes from a debate we entered about the character of Indigenous knowledge of undersea worlds, especially with respect to corals and how they were used. Archives from the 1860s spoke of people in the South Pacific islands who had developed deep knowledge of coral undersea gardens and the ecological interactions in them, and moved and 'planted' corals; others were sceptical of the extent of this underwater knowledge, however, suggesting that it must have been restricted without glass mask technologies that developed only from the 1920s. Yet those who spend time under water can adapt their sight systems to see very clearly, compensating for disruptive underwater parallax. In effect, people are able, through experience, to come to operate in an undersea *umwelt*, aligning their sensorial worlds through repeated practice and time spent living there. The glass of goggles, when they arrived, opened up underwater worlds to a wider human gaze in a way not dissimilar to the glass of our television screens. While it democratised the gaze to many more people, it involved a more distanced and mediated form of sensory alignment, one less dependent on (and productive of) intimate and prolonged underwater life. Nevertheless, there is no doubt that a glass mask-mediated gaze can engender appreciation of and affective engagement with sea assemblages, just as many people become emotionally affected, often deeply, by the narratives of underwater documentaries.

Sea–Human Lives, Negotiation and Respect

In people's engagement—or co-becoming—with sea assemblages, affective dimensions are often entwined with making a living, or with other human social purposes of sport or leisure. We saw this with regard to the sport/life of surfing, but it is relevant under the water too, including to fishing in its many forms. How people engage with animate sea assemblages in such contexts is again shaped by how they frame them; on the kinds of life, beings, agency and intentionality they understand to be in play. Amongst many Māori, for instance, seascapes and their creatures are considered ancestors (*tupuna*) and kin (*whanau*), and they are all imbued with *mana* (a sort of spiritual and physical authority and power) and *mauri* (a spiritual life essence). Reciprocity of respect is essential and blessings (*karakia*) are spoken before taking fish from the sea.[24] Amongst Yawuru people, the original owners of Broome in coastal Western Australia, the whole seascape and its more-than-human inhabitants are included within conceptions of 'country' and 'culture', within which maintaining frequent connections, and maintaining and fulfilling relations of responsibility and respect, are central to *mabu liyan* or good *liyan*—living well, including a sense of belonging, emotional strength and pride.[25] People in Tanzania's Mafia Island who live from fishing notice and ascribe characters, even personalities, to different kinds of sea creature. Everyday fishing experiences are intimate ones, and stories of seafaring describe these.[26] In many examples shared by anthropologist Veronica Strang, sea assemblages include personified water serpent beings or deities with powers that require respect and careful negotiation.[27] For instance, Inuit in Greenland have described the western Arctic seas as the home of Sedna, a powerful sea goddess with a mermaid-like form who rules the underwater spirit world and is widely regarded as the mother of all other living sea-beings, which are continually made from her body, and who demands mutual respect. Hunting of sea mammals such as whales, walruses and dolphins is

> a sort of social transaction between humans, the souls of sea mammals and Sedna [...]. A cyclical flow characterises this transaction in which the Inuit receive the bodies of sea mammals whose souls then return to Sedna to be regenerated or reproduced. Humans may negatively affect this cycle by breaking certain taboos, in which case Sedna responds by cutting off the food supply or otherwise causing trouble.[28]

What is often going on in engagement between people and seas, it seems, are matters of negotiation; of conversation; of wits. This is negotiation not just with a single being, but with assemblages, and humans participate not as external communicants with, but as members of, a more-than-human communicative community. Thus the wise Inuit hunter avoids the transgressions—hunting out of season, or hunting sea and land mammals at the same time—that will anger Sedna, or must bring in a shaman to propitiate her and restore order. The surfer who seeks to 'catch a wave' and bring their board in securely and elegantly is engaged in a game of wits with the sea's elements. Surfers often read the sea as a capricious world. Seeking the big wave is about putting oneself in the ultimate position of vulnerability with respect to the sea's unpredictable forces, and then seeking not to control them, but to outwit them. The accounts of people who live from fishing, as in Mafia Island, describe it as a matter of wits, almost like a game in which one is trying to outwit the fish and the elements it swims in, in which it always has a chance to escape. Fishing for sport is like this too; and it was the way the Sierra Leonean women with whom Melissa (writing here) used to net-fish in streams described what they did.[29] You can be helped by knowing your fish: where it is likely to be or to hide, when in the day you are most likely to find it, how deep to go with your rod or net. Fish vary widely, and have different species and place characteristics, sometimes even different personalities. It is only through outwitting them that one can succeed. This requires us to endeavour to 'think like a fish'—a kind of mimesis that is the opposite of anthropomorphism, which would involve thinking that fish think like us. And it is a matter of thinking, too, like a sea assemblage, of which fish and human are co-members.

Fishing is usually about killing fish, to be sure. But respect also comes into play. Just as land-based hunters in many societies know and seek the respect of animals when hunting, likewise expert marine hunters and fishers often do so. We have already alluded to examples from Greenland, Tanzania and Australia. In Sierra Leone, women were adamant that some technologies created an unfair balance of power, a lack of respect, between humans and fish; they were very cautious in their use of herbal fish poisons, allowing only the most experienced older women to use them, and deplored the dynamite they had heard of being used in some larger rivers. Some catfish and carp in the West African world are never hunted, as they embody ancestors.

This is all very different from the world of mass industrial fishing, which engages in sea assemblages in very different ways. Here sea assemblages and their communities of communicative, animate life are reframed as 'fisheries'. Powered by the force of commerce and enabled by the technologies of large boats, engines, scanners and nets, fish are scooped out of the sea indiscriminately. The elements are overridden. The intricacies of networks amongst fish and other creatures are wiped out, replaced by the encompassing categories of 'catch' and 'bycatch'. These are the terms that dominate the language of policy in what is in effect the abbatoir of the sea. This is all very different from engaging with a communicative sea and its animals in a game of wit and mutual respect. A high modernist fishery is still an assemblage, but of a different kind, imbricating multiple beings and technologies in different ways.[30]

Marine conservation initiatives, likewise, often overlook the power of engaging with seas as communicative, more-than-human assemblages. Those involved conceive either that they are conserving 'marine ecosystems' imaged in mechanomorphic terms, or that they are bringing seas epitomised as 'wild nature' under containment, leading to what Gisli Pálsson, drawing on Michel Foucault, calls 'the birth of the aquarium': the rise of management regimes that treat the sea as a mighty aquarium that must be enclosed.[31] As Stefan Helmreich points out, such enclosure is predicated on a separation of nature and culture, with the sea scripted as a hypernature until now outside culture.[32] Such views are contradicted by many maritime anthropologies,[33] and also by Indigenous accounts that reveal people's framings of seas as lively, animate and interconnected with human worlds. They are contradicted still further by the more-than-human ethnographies viewed through a structural-biosemiotic lens that we have developed here, which reveal sea assemblages to be communicative communities uniting human and non-human members of naturekind.

So while whales, dolphins and octopuses may be poster children for advances in understandings of animal intelligence, culture and language of a certain sort, a structural-biosemiotic framework extends further, revealing a wider language of sea assemblages. By taking the focus off these charismatic species and off the studied association between their large brains and linguistic powers, our structural framework allows appreciation of the myriad ways in which other sea beings communicate, with each other and with us, in ways that depend on structured grammars and signs in fields of signs. All animate sea beings participate, as do inanimate ones as well, in more-than-human sea assemblages that are thus revealed as held together by webs of structured semiosis. This is very different from the kind of mechanomorphism found in portrayals of marine ecosystems. It sharply contradicts the approaches characteristic of policy debates that reduce sea creatures to resources and commodities, and emplaced habitats to fisheries. And it adds to concerns about the 'shifting baselines' that marine ecologists have observed, whereby, as numbers and diversity in marine life disappear, we recalibrate to forget this.[34] What is lost is more than sheer numbers; it is historical memory of a rich more-than-human communicative assemblage, and opportunities to be immersed in it into the future, with all the affect and mutual flourishing that this can entail.

11
Cities

We want to avoid any implication that our analysis of language and culture beyond the human is somehow 'rustic' and of limited, or declining, relevance to a world in which 57 per cent of people, expected to rise to at least 68 per cent by 2050, are already urban dwellers,[1] or that cities are devoid of non-human nature. On the contrary, towns and cities as assemblages usually include an enormous diversity of non-human animal and plant life that interacts with human inhabitants in a vast variety of ways, even if these are sometimes not immediately obvious. Our aim in this chapter is to probe the communicative dimensions to such interaction, and their significances for the life, health and well-being of people and other beings dwelling in cities: that is, in urban naturekind.

We develop three particular lines of argument. First, we consider how a city, as a whole, is in some way communicative, and probe how a structural-biosemiotic approach reframes how this has been construed in urban studies. Second, we again consider the lively interplays between people, the animate and inanimate, and how viewing these through a structural-biosemiotic lens reveals the conversations and communicative communities involved, and what might be distinctively due to their urban emplacement. Third, we consider implications for human health, particularly mental health, and how a structural-biosemiotic approach might add to existing understandings of how 'green spaces' and experiences of connection with non-human nature in urban areas have therapeutic qualities.

Communicative Cities

Despite the diversity of cities, there is something distinct about town or city assemblages in general: a grammar to the urban that is different from the non-urban. Currently the world has forty-four megacities (each with a population of at least ten million), and ninety-seven of at least five million human residents. A smaller settlement is usually defined as a city (not a town or a village) if more than a hundred thousand people live there. Cities are often framed by inhabitants and planners in mechanical ways. More recently some have come to frame cities as 'living', whether as 'like organisms', or as living assemblages forged through myriad interactions between human and non-human beings and entities.[2] Taking a communication approach reveals cities not just as lively, or comprised of lively interactions, but as more conversational: places where communities are created through the sharing of meanings.

Some framings that consider cities as 'living' suggest that their multiple components are linked through complex interrelationships and feedback loops in ways analogous to a living organism. This is captured in emerging studies of 'urban metabolism',[3] and as a common vision in advocacy movements for urban sustainability. Thus, as Anders Lisdorf, writing on the 'people-powered solutions' network *Shareable* argues,

> [m]odern physicists and city planners are now beginning to argue that rather than viewing the city as a conveyor belt, it should be viewed as an organism. Organisms have a metabolism that converts energy and resources into the nutrients and substances that they need. The same is the case for cities. They also take in energy, raw materials, and water and convert them into heating, entertainment, houses, and pretty much everything we associate with living in a city. Just like a person's metabolism, a city's metabolism determines how well resources are used and stored, which in turn impacts a city's health and the surrounding environment. Consequently, the metabolism of cities is an important area of focus when it comes to optimizing the use of energy, water, and raw materials. By better understanding how cities work, it becomes possible to apply insights from the circular economy, which focuses on reusing and re-purposing waste products as resources.[4]

Such a conception, of a city as living, alters approaches to urban planning. The US advocacy organisation StrongTowns, for instance, argues that living cities are constantly evolving, potentially optimising to the needs of the people from the bottom up. Conceiving of them in this way enables planners to accept plenty of small-scale chaos (trial and error) at the local level, loosening restriction on property use, allowing a more emergent approach, and planning for a high level of incrementalism (planning buildings and infrastructure that can be adapted and added to over the generations as their needs change even in unforeseen ways). These metabolic framings thus reconceive urban systems in ways that capture the pulsation of city life and materiality, and the unfolding complexities of their evolution—the vitality of a city as living, in contrast to more rigid planning—even though such approaches are more the exception than the rule today, as modern urban planning generally favours their mechanical opposite.[5]

These metabolic framings focus on cities as 'wholes': as more than the sums of their parts, growing, living and adapting in much the same way as a growing organism. Other approaches are more focused on the component parts and processes, human and beyond, that contribute to urban assemblages, albeit remaining attentive to the emergent properties of cities as living achievements.[6]

Our own approach aligns with works that take posthuman and multispecies ethnographic approaches to urban settings, revealing them as composed of lively interactions between a multitude of human and non-human entities, while attending more fully than others have done to the communicative dimensions of these. For instance, vital materialist framings, whereby all objects are understood as lively and communicative, have been applied to urban settings. The emblematic starting point for vital materialist philosopher Jane Bennett is a grouping of objects around an urban drain—a plastic work glove, a dead rat, a stick of wood, a bottle cap—which convey meaning to her, too many meanings to pin down. As she writes, together they 'issued a call'—they were noticed for some reason: for 'the contingent tableau that they formed with each other, with the street, with the weather that morning, with me'. Each thing had its own vitality, as 'vivid entities not entirely reducible to the context in which (human) subjects set them, never entirely exhausted by their semiotics'.[7] Reconfiguring her analysis through a more structural lens we can note that each of these objects certainly carried meanings of its own (in relation to fields of other signs, albeit subliminal to her), and yet they carried greater meaning together, in their juxtaposition, though in ways that could not be entirely pinned down, given the infinite multiplicity of possibilities. It was the combination that spoke, which itself thus acquired higher-level meaning as a metasign (for her), including of waste, and thence of hyperconsumptive materialism in America; acquiring sense, then, in relation to narrative structures (metasyntax) of which they are a part. They also hinted at other possible meanings, thus ensuring that none was authoritative or stable. What is important is that while the vitality of objects might not be 'exhausted by their semiotics' it cannot be dissociated from these semiotics.

Considering urban material objects as vital and communicative in such ways, as components of, and emergent from, a lively assemblage, is rather different from work on urban networks adopting actor–network perspectives, in which—for example—the fabrics and energy flows in a city (in a commercial office building, say) are conceived of as 'actants' (not communicants, or even communicators).[8] Debates over whether and to what extent urban material objects and assemblages can or should be treated as having life and agency—even cognition or intelligence—are by and large irresoluble, invoking, as they do, framings that vary enormously across the pluriverse. Vital materialist framings, wherein all objects are considered lively and agentive, despite lacking animate life or cognition, offer one set of resolutions. In other quarters of the pluriverse, such questions would be resolved differently: for instance, by giving urban rivers the status of persons. Applying our structural framework, however, we can, as we have been arguing, take the focus off these questions, allowing us to consider how non-living and living beings alike communicate with us in a context of structured fields of signs, grammars and packaged meanings. In this way we can understand how all inhabitants codevelop language in and with cities, and consider the power of that language—in debates over urban planning or sustainability, for instance—without having to resort to analogical reasoning. A focus on cities, where the communicative aspects of the material fabric are so pronounced, helps us to realise that a structural biosemiotics does not have to end with the 'bio', but can also encompass entities that are not part of biological life as it is conceived, at least, in scientific orthodoxy.

Urban language does not just encompass micro-communications amongst cohabitants, but extends also to wider spatial and temporal signs and grammars, metasigns and narratives. Thus for those who know a given city, the layouts and architecture of municipal quarters or neighbourhoods make sense in relation to others, with their juxtapositions conveying to us social and economic information. Our day-to-day navigation of urban quarters includes observation of such orders, and of matter out of place in them—a shiny luxury goods shop in a quarter that otherwise conveys impoverishment, perhaps—as well as the sensation of ourselves being matter out of place, possibly, in areas of affluence and surveillance. Such experiences allow creative meanings: of knowing our place, or riffing on it, perhaps by graffiti tagging; by asserting presence and inscribing meaning.

The temporalities of city life have orderly rhythms, often shifting from early morning quiet and emptiness, through the thronging streets of rush-hour and a working day, into nightlife; as well as seasonal rhythms: social, climatic, economic. The signs evoked by buildings, their textures and scents, all speak not just of the present, but of histories—of personal experiential or learned histories that shape the grammars, metasigns and narratives (metasyntax) of encounter. The empty plinth by the docks in the British city of Bristol where once there stood the statue of slave trader Edward Colston could, for instance, be considered a metasign: a single object that, perceived in relation to others, now packages a complicated centuries-long history of city construction and prosperity built on the buying and selling of people, the struggles against that trade, and, in the 2020s, the toppling of the statue into the water amidst contemporary anti-racist struggles: a rejection of the proud public display of the city's past, and a counter to nationalist, racist narratives.

Urban Lives Beyond the Human and their Structural Biosemiotics

Cities are assemblages of more-than-human life, encompassing a multitude of plants and animals. It is in acknowledgement of this that geographers Steve Hinchliffe and Sarah Whatmore refer to 'living cities',[9] with emphases rather different from those of the vital materialists, attending to how humans and urban infrastructure mingle with green spaces and non-human species in recombinant urban ecologies that are crucial to what makes cities liveable for humans and non-humans alike. Green spaces include private gardens, roofspaces and the greenery growing up the outer walls of buildings of 'biophilic' design, leisure spaces and allotments, feral spaces of railway sidings and derelict land, and remnant spaces of waterways and woodlands. Such urban ecologies and their biological diversity blur boundaries between the domesticated and the wild, but are recombinant because, as non-human and human beings alike adapt to urban life, this shapes their co-being—and, we would stress, their communication.

Green public spaces convey meanings as part of urban grammars within a whole variety of ranges. They acquire sense in relation to grey and brown, for example; they are often public, not private. They convey packaged meaning in relation to other metasigns: as places for leisure in contradistinction to work, shopping or transport; as offering the potential for tranquillity rather than stress; as spaces for stillness, not movement; and so on. These are contrasts that acquire significance both for people and, we can infer, for the non-humans who frequent them as part of their fields of spatial meaning. Immersion in a green space in a city, whether resting in a park at lunchtime, walking a green path to work, taking a dog for exercise or spending time in a garden or allotment, is significant at least partly for its contrast with the urban space, life and rhythms around: it is time and a place apart, a break from all that. Whilst the signs of being in a green space might be felt as embodied, just as we feel an insult or compliment bodily, and blush, such signs acquire their meanings, as words do, within subliminal languages. The sense of place (and of being out of place) is felt; it is emotional—but not prelinguistic. Importantly, heightened significance (and emotional affect and effect) might accrue to cultivation of what is more or less the same kind of allotment in an urban as opposed to a rural setting, where the ranges of meaning produced are different (stereotypically, less extensive: for rural dwellers, digging in a garden or a walk in a park provides less of a contrast with the surroundings; a less distinct sign or metasign). It is this structural context that is not captured by approaches that stop at considerations of embodiment. The structural-communicative aspect of green space has been given relatively little attention, and perhaps deserves more.

The lives of non-human urban inhabitants extend beyond the boundaries of any recognised or designated green space to encompass life and interactions in and around streets and buildings, above them around rooftops, and below in cellars and sewage pipes. From the plants that settle and creep through cracks in brickwork or where pavements meet walls, to the birds and insects that nest and fly around roofs, to the rodents that occupy their underworlds, cities are vast assemblages of life. And some of these creatures and plants are there, of course, because humans have introduced them: urban beekeeping, pets and companion animals who roam the streets, trees and plants in public parks and people's gardens. This array and its webs of interconnection are increasingly studied by urban ecologists. Zoologists and biologists are interested in how species adapt their habits to urban environments, and in the overlap between urban wildlife and domesticated species, as creatures cohabiting cities with humans develop particular ways of being. As some social scientists now attend to the more-than-human interactions that go on throughout urban assemblages, making city life an interspecies experience for humans and non-humans alike, we can add to this by engaging with the urban biological revolution to consider how it is communicative, and in structured ways.

Animals and plants in cities communicate with each other (and with humans) through their repertoire of senses. Whilst their perceptive world—their *umwelt*—is characterised by some biosemioticians in relation to their sensorium, questions arise over whether this is a matter of species, of the individual, or indeed of something altogether more emplaced, in which varying sensorial possibilities and significances become manifest. Looking at how the communicative dimension of life transforms in cities might suggest the last of these. What is significant to members of the same species may be quite different in rural and in urban settings, with communication adapted accordingly.

Urban pigeons provide an example. Pigeons are part of many city assemblages, dwelling on the roofs and ledges of urban buildings and in the trees in streets and parks. The wild rock doves from which they descend historically bred on cliffs in colonies,[10] and this evolutionary origin helps explain why urban pigeons look for vertical and rough surfaces for breeding. The architecture of some old buildings provides a profuse range of nesting opportunities,[11] whilst the birds find abundant food in the waste tips, bins and surroundings of human homes, shops and restaurants, reworking their feeding behaviour and movements accordingly.

Urban pigeons may also have adapted their navigation systems. Studies of homing pigeons have revealed them to have sophisticated abilities to navigate across long distances. This is based on three navigational strategies, mobilising three different sensorial modes that shape the pigeons' *umwelt* in relation to the world around: magnetism, in relation to the earth's magnetic field (a magnetic compass); a sense of time, in relation to the angle and polarisation of the sun's rays (a sun compass); and smell, in relation to an odour map.[12] Whilst each of these can be demonstrated by experiments in animal cognition, how meaning is acquired from them and their integration is more assumed than theorised. Yet it would be possible to probe these as forms of pigeon navigational language. Each of the navigational systems is 'structural' insofar as the signs picked up on make sense in relation to established orderings: urban magnetic signals, solar rhythm, the emplaced contours of odour and so on, with each acquiring packaged meaning related to experience. What is interesting is that even if one navigational system fails, it appears that pigeons are able to mobilise one of the others, and not be thrown off course. This ability to integrate navigational strategies and shift flexibly between them has been identified particularly amongst older homing pigeons, and seems therefore to be a matter of accumulated experience and learning.[13] Whether in the case of individual pigeons or of groups, one might reflect that some amongst these strategies would be especially appropriate to navigating in an urban area, and that the ability to switch between them could be very valuable in city life—where high buildings might block sunlight, for instance, or traffic or industrial fumes disrupt the smellscape. One might even speculate that navigational switching has been integral to the co-evolution of pigeons with urban spaces and their capacity to flourish there. Given that these birds have powerful learning abilities—they can learn to distinguish Impressionist from Cubist art when to do so is made relevant to them, for example[14]—we might well presume that they learn to differentiate what matters to them, picking up the meanings conveyed by signs and their juxtapositions. Such questions of pigeon language and its urban versions, or dialects, thus must extend to the question of pigeon culture, and of how learning and experience become shared amongst groups of pigeons differently according to the specificities of cities and the experience of dwelling within them. Culture of this kind emerges not only as each pigeon acquires common emplaced experience, but also as pigeons relate to each other in the collective development of such emplaced experience.

Communication between humans and pigeons in cities can involve the fleeting interactions that happen through encounters in the street or a park. They can involve more intense kinds of communication: for instance, when pigeons choose to feed on allotment or garden plants, and people's reactions may be to shoo them away with shouts and waves, or to erect screens and barriers, limiting the conversation in the interests of keeping garden plants alive. Urban pigeons are commonly depicted as 'pests' in most major European cities, much as seagulls are, and have been the focus of personal and institutional efforts to control their abundance. More recently, the sense of threat has extended to the viruses and other diseases pigeons might carry, viewing them on this account as a menace to humans. The observation that '[w]hen animals colonize new niches by adapting to and settling in places different from their original ecosystems, they may present a visible contradiction to human definitions of these landscapes and their assigned species'[15] also begins to suggest the idea of pigeons being structurally polluting. Cities are subject to historically embedded and emplaced ordering processes—grammars—that shape what is deemed appropriate in belonging to them. In the case of pigeons (and seagulls), dominant human framings have excluded them from the appropriate order of the city. They are labelled as dirty, nuisance species: pollution, that is, or matter out of place.

Yet human–pigeon relationships in some cities involve more intimate companionships, that shape the experience of city life and assemblages for both. In many towns and cities in the Punjab areas of Pakistan, for example, rooftops are dwelling places for pigeons that men often keep for companionship and leisure. Pigeon lives are ordered by their rooftop coops and the flying sojourns they undertake at the behest of their human companions, who form an intimacy with them, 'feeding them expensive seeds and herbs, decorating their feet with anklets and bells, and flying them with gusto'.[16] As anthropologist Muhammad Kavesh recounts, the experience of one such pigeon-flyer captured well the intense multisensory communication involved in establishing and maintaining this more-than-human order:

> Because [he] spent extended periods of time with his pigeons, he was able to form a sensory relationship with the birds. The texture of pigeons' droppings informed him about the birds' feelings, moods and level of contentment. By monitoring pigeons' flight patterns, he could tell which type of training they needed most. By observing the birds' breathing rhythms and blinking frequency, he could pinpoint their fears, issues with a mating partner and satisfaction in their designated coop. His vast knowledge of pigeons' diets enabled him to select a different combination of seeds, nuts and herbs to support the pigeons' digestion, increase their endurance and stamina, and make the birds forget their thirst in June's unbearable heat.[17]

For the pigeon-flyer, the intimate, joyful connections offer some alternative to and respite from the hardships of everyday life—as this man put it, with his pigeons his 'inner-self becomes happy'. This affect, however, emerges out of interspecies communication rooted in the codevelopment of grammars and fields of signs, enriched through experience; a language that is not simply a human one but also a pigeon one, emplaced within an urban assemblage. It is impossible (and unnecessary) to gauge what the pigeons make of it, but the ordered way they return from flying suggests that there is something attractive—affective—for them too in this very particular urban way of life. It might relate partly to food, yet if so, this should not be construed mechanically, in relation to instinct, but in relation to how food, pigeons and handler are linguistically emplaced amongst the rhythms of flying and rooftops.

Urban monkeys provide another example of ways in which animal communication is reworked in city life, and becomes part of interspecies urban conversations. A detailed multispecies ethnography by Maan Barua and colleagues provides a basis to appreciate this, but also to tease out the potential for a structural-biosemiotic framing.[18] They describe how rhesus macaques have become ubiquitous in Delhi and other northern Indian cities over the last four decades, driven by politico-economic forces, including capture for commercial trade and export, and then abandonment as exports were banned. As macaques have become urban, they have shifted to anthropogenic food, obtained either as people feed them, or by salvaging waste or raiding people's homes. They share places frequented by humans, from parks to temples. People have fostered this through everyday practices. Thus 'Delhi witnesses large-scale feeding of macaques, from passers-by, pausing to buy bananas from street vendors, to the affluent middle class, who bring food in cars, strewing large quantities of grams, vegetables and fruit by the pavement, thronged by macaques'.[19]

Interspecies communication has now become central to the lives of the macaques and humans alike. It involves sound, presence and movement, combined in distinct, emplaced grammars: those of the street, and those of the temple, for instance, the latter shaped by the daily rhythms of people's visits and the more continuous presence of the macaques. In this way humans and macaques share and exchange meaning, developing conversations that are central to city life for both. In their conversations, however, people read the presence of macaques through plural framings that envisage their place in urban orders in distinct, if sometimes overlapping ways. Thus in one narrative framing, people understand the lives of macaques and their place in urban assemblages and habits around temples as similar to those of people, in their sharing of space and food. This coexists with Brahmanical framings through which people envisage macaques as intermediaries with deities (especially the deity Hanuman, an incarnation of monkeys), and feeding them then becomes a way for a person to harness *punya*—a diffuse, cleansing merit. There is the possibility, too, for astrological framings, astrology being both a Brahmanical, scripturally sanctioned practice and a growing commercial business in Delhi, whereby provisioning macaques can be a remedy for adverse social and ecological effects wreaked by the planets, such as ruptures to kinship. In this way the pleasures of feeding macaques are linked to ways to repair disruptions to urban orders that extend to the cosmological. Those who visit and feed temple macaques switch between such framings quite comfortably; moreover, whilst these are framings constructed by humans, macaques are participants in them. The macaques are not simply objectified by people in these discursive framings or narratives, but pick up on and respond to the more-than-human regularities shaped by them. The codevelopment of interspecies languages is thus inseparable from human cultural practice. That temple macaques and street macaques have such distinctive behaviours suggests, however, that it is equally inseparable from macaque cultural practice. It is 'interspecies intercultural'; a further example of what Lestel refers to as 'hybrid cultures'.

Cities, Communication and Health

A third implication of taking a communicative lens to more-than-human relations in cities relates to questions of mental health and well-being, and the significance for this of communication beyond the human.

Whilst there is a long-established literature on the social and biological determinants of mental health,[20] an emerging literature exposes the importance to human health, and in particular to mental health, of interactions with non-human nature in cities. This generally deals with questions of access to green space and improving it in planning, on 'biophilic' design that creates opportunities for interactions with non-human life, and in some settings, the emergence of 'green prescription' as a biomedical treatment.[21] Might a focus on more-than-human communication have a contribution to make here, helping to understand such encounters and their health implications, and to identify the qualities of such interactions that might be significant to them?

From the perspective of human experience, academics and artists alike have long elaborated on the distinctiveness and significance of urban life. Sometimes this is in a positive vein, with city life experienced and imagined as replete with opportunity, excitement and progress. Others highlight more negative experiences: cities as theatres of failed hopes and dreams, social alienation and exclusion. Unemployment, homelessness and distress are all part of the contemporary dynamic of capitalist urban life, experienced in highly unequal ways. That urban spaces generate mental distress, be it through social segregation, inequality or violence, has become a refrain in the social sciences.[22] Medical scientists and practitioners, too, pay increasing attention to the prevalence of mental health problems amongst urban inhabitants, alongside physical health problems and inequalities, including those linked to pollution and waste, to disease prevalence and transmission and to living conditions. Some note a potential urban bias in such work, as negative social experience and anxieties of the same type affect those living in rural worlds, too, with other literatures identifying rural disadvantage, policy marginalisation, prejudice and stigma.[23] The differential between urban and rural settings in terms of proximity and intensity of engagement with more-than-human life has barely been factored into these debates about urban and rural mental health, and deserves more attention.

Numerous studies associate more access to green space in cities, including parks and gardens, public and private, with better mental health. In an international review of 263 studies relating to green space and mental health, 70 per cent of them reported a positive association, although not all had corrected for other social and economic attributes that come with such proximity.[24] In New Zealand, proximity to green spaces is found to be associated with reduced anxiety and mood disorders.[25] A cross-sectional study of four European cities found links between time spent purposefully in green spaces and improved levels of well-being and vitality.[26] Reported benefits include both short term spikes in well-being and the potential to increase resilience against a stressful life. Relatedly, 'green prescribing' has become part of the health policy lexicon, advocating (and sometimes making provision for) 'doses of nature' as a route to tackle a range of physical and mental conditions. This is proffered as a cost-effective substitute for, or supplement to, other modern biomedical interventions.[27] One study from England suggests two hours of 'recreational nature contact' weekly as an appropriate 'nature dose'.[28]

But what is actually going on when people take a 'dose of nature'? How do these mental and emotional effects work, and through what kinds of affect? Knowledge of the benefits of our more-than-human experience to our health is largely premised on association—on correlation—and an intuitive understanding of cause, and when more causal explanations are evoked linking surroundings with emotional and mental state, authors are forced to deploy a lexicon of terms such as 'visceral' and 'embodiment', that never quite fulfil their users' explanatory ambitions. More-than-human ethnographic studies, and attention to more-than-human communication, add some valuable insights.

As we have seen, encounters with pigeons on a roof, or in a park, can produce positive feelings, whether of relief and reassurance, or distraction from other stresses. Such affect emerges through communicative encounter, through codeveloping a language and conversation. Thus Barua and colleagues' informants in Delhi, for example, reported ways in which their communication with macaques affected their emotional well-being, as in the case of Anuj, who found a way of dealing with the cumulative anger and sadness he felt in a difficult urban landscape through everyday acts of fasting himself, whilst feeding macaques; or Kusum, whose feeding, sound and gestural exchanges with the Hanuman temple macaques, interspersed with prayer rituals, brought her 'peace and harmony' and enabled her to deal with family predicaments: "My two sons are now doing well, and it is entirely because of these rituals."[29] Such affective entanglements thus give rise to what Barua and colleagues term 'microspaces of wellbeing' within wider more-than-human urban assemblages, with the potential to alter experiences and conditions of ill-being and immiseration, even if in small ways. Such practices, which have been termed 'niches',[30] but that we might characterise as 'emplaced', enable people to render city assemblages more viable and habitable; yet they 'are not the domain of the human alone. The rhesus macaque too constructs niches. [...] Macaques' modes of existence thus intermesh with how people relate to the city and have bearings on the experiences, lives and wellbeing of both.'[31] In a further example, the Ebony Horse Club in Brixton, London, enables young people from socially disadvantaged backgrounds to spend time engaged in a variety of human–horse interactions that it is hoped will improve their mental well-being, as well as life skills, education and aspirations. The club's yard, surrounded by busy streets and housing blocks, provides a distinct space in the wider urban assemblage in which emplaced human–horse communicative encounters can unfold in ways that bring relief and escape from the stresses of these young people's urban lives.[32] Events that might seem banal to those living rural equestrian worlds become therapeutic when emplaced in city assemblages.

In such examples, therapeutic qualities turn at least in part on experience and communicative encounter beyond the human, suggestive in itself of what might be missing from everyday life. What is true for individuals in such examples may also be the case on a wider scale, as communicative connections with non-human nature provide therapeutic relief and possibilities of hope amidst the annihilation of the world around—a point we explore further in the book's conclusion. Meanwhile, these examples, viewed through a communication paradigm, suggest that in city assemblages several factors may be at work in therapeutic affect. First, those suffering stress and manifestations of mental 'unwell-being' might find benefit from the creativity and intimacy of codeveloping for themselves more-than-human languages. Second, there are distinctive routines and rhythmic orderings to such encounters that are distinct from stressful everyday city life. And third, such encounters are differently emplaced—the rooftop, or city riding yard, or temple contrasting with the wider city assemblage around. In short, affect, and mental health effects, are associated with more-than-human communicative experiences that are distinct, temporally or spatially, from the standard run of challenging, everyday city life.

Such experiences might also manifest in less distinct ways, more threaded through everyday urban life. Such is the implication of an aptly entitled study of 'the magic of the mundane' in Sheffield in the north of England, whose insights again can be considered through a structural-biosemiotic lens. Here, Dobson and colleagues revealed (unsurprising) relationships between better general health and larger average garden size, greater total green space cover and greater local tree density.[33] They also found lower levels of depression in areas where average garden sizes were larger and where publicly accessible green spaces were cleaner. Yet they consistently found too, in people's own accounts, 'the effects of routine or incidental encounters with urban nature in enhancing individuals' wellbeing and outlook on life'.[34] People noted everyday experiences that made them feel better, referring especially to their 'wonder' at encountering wildlife and enjoying birdsong, and seeing trees and vegetation and noting how it changed through the seasons, especially on journeys to work; and to the 'awe' evoked by colourful, dramatic skies and views across the city. Karen, for example, described the scrubland seen on her daily journey as 'always different. It ebbs and flows like the sea.' In some cases, mundane experiences such as sitting under a tree, watching leaves in a park or holding a twig in a pocket offered opportunities for self-reflection on 'life problems' and to gain a sense of 'what is real' and manageable. This is not obviously communication, yet the twig in a pocket conveys a whole package of meanings, from the flourishing of urban trees with all that they signify, to one's connectedness with them. As a sign, it has iconic qualities, in that the twig stands for the tree and the land it grows in; at the same time, however, it has symbolic qualities referencing experiential encounter, the moment, the memory. One could reflect further that wonder at encountering birds and animals, and the feelings this evokes, might relate to their positioning as matter out of place in city assemblages otherwise dominated by humans and their infrastructure. Yet such 'out of placeness' would depend on Cartesian framings that divide (urban) culture from nature, and which are erased by so much evidence and experience that cities are more-than-human assemblages, replete with life of all kinds. Dobson and colleagues emphasise the importance of such everyday and momentary experiences for urban mental health, and that such are experiences are consistently undervalued in discourses of 'green spaces', 'green prescriptions' and 'nature doses' that assume connections with non-human nature to require planned and designated spaces and times, paying more attention to their quantity than to what actually happens in them.

Exploring cities thus reveals disjunctures between framings and discourses in which the urban epitomises human orders, associated with mechanistic modernity, progress and the vibrancy of capitalist commerce, and those in which the urban consists of vibrant assemblages, ecologies, of many more-than-human lives and ways of being. The former have long dominated urban planning and governance regimes and their underpinning sciences, and have gained recent ground in discourses that promote 'smart cities' as harbingers of the latest (including 'sustainable') infrastructure and technologies, but as largely devoid of non-human life. Yet as the implications of such discourses are increasingly felt, in marginalisation of the people who fail to fit the planning framework, in the unsustainability of so-called smart systems, and in stress and ill-health of humans and the remnants of non-human nature surviving in cities; so greater appreciation of cities as living assemblages, with all the benefits this can bring to human and non-human nature, is resurfacing. In orienting ourselves towards living, not dead, cities—towards cities conceived as assemblages of and for naturekind—recognition and support for green spaces, urban biodiversity and the promotion of people's interaction with these ('doses of nature', in other words) are certainly to be welcomed. But they can be enriched by a deeper, more textured understanding of more-than-human interactions within and beyond these spaces and moments; interactions which are intensely communicative, and work in structured ways to shape languages in and of cities. As for the other assemblages that we have considered, attention to everyday communicative experiences of cohabitants—experiences often obliterated by dominant discourses—brings fresh insights into how such languages operate, and why they are important.

12

Conclusion

A POLITICAL NATUREKIND

This book has addressed the impasse between the accusations of anthropomorphism that natural scientists often throw at those in the humanities who discern intersubjective experience with beings beyond the human, and the accusations of mechanomorphism that those in the humanities throw back at natural scientists who have found meaning superfluous to parsimonious explanation. This is a crucial impasse for our era. We face a crisis of the annihilation of naturekind and its living assemblages on which we depend. But are these assemblages comprised of mechanical interactions amongst actants alone, or also of conversations amongst communicants? If the latter, it is not just the extermination of all that once existed that should concern us, but also the annihilation of our communicative and emotional encounters with what remains; of our kindred and sociality beyond the human that is so important in making us care.

This book has built on new alignments that have become possible: between the biological revolution in the natural sciences that reveals social learning, cultural diversity and linguistic communicative practices across naturekind almost wherever it looks, and the post-human revolution in what used to be the 'social' sciences and 'humanities' but that no longer restrict social or semiotic theory to people and that no longer restrict reflection on this to academic ivory towers.

Yet whilst these new alignments enable us to recognise afresh the significance of communicative encounter beyond the human, they have been less forthcoming on exactly what this might entail; on what it must look like, and why. Whilst many have argued that answers will lie in probing communication 'beyond language' or 'before language', we find the answer to lie instead in taking language itself 'beyond': in expanding the insights of structural linguistics and semiotics far beyond what has hitherto been construed as language, in the approach we cast as 'structural biosemiotics', and exploiting the broader conception of language that this allows.

Having laid out and illustrated this new paradigm, in this final chapter we now revisit with more clarity the plethora of wider separations (or alienations) that interlock to divide us from communicative and emotional entanglements with non-human natures, ranging from things physical to those that are more conceptual. A focus on communication beyond the human casts new light on how such separations come about and why they matter so much, and also on their limits. This opens up important possibilities for recoverability: for finding more convivial relations than heretofore across naturekind and for configuring approaches to care and justice. Here we build in association with others contemplating a more-than-human justice and politics, considering the implications of a structural-biosemiotic framing for the scope of environmental politics and what it involves, and for the enactment of such a politics, taking inspiration from parts of the pluriverse where communication beyond the human is more generally accepted and appreciated. There are implications, too, for discourses and practices of biodiversity conservation and restoration, and for practical efforts to build sustainable futures on the planet we share with all life.

Communication Beyond the Human and the Power of Meaning

The philosophies and sciences that bolster notions of human exceptionalism, and the hierarchies, anthropocentrism and speciesism that so often follow from these, have proceeded on the basis of a fundamental presupposition that the capacity for language is uniquely human. Yet the more one probes all the dimensions of communication, the less significant the version of 'language' implied by this becomes. When language is reconceptualised in a broader way, as multisensory, structured communication, the core division between humans and the rest crumbles, and with it much of the basis for conceiving of humans as so fundamentally different in other ways too. The particular aspects of language, narrowly conceived, that do seem to be uniquely human, including the particularities of speech and symbolic writing, come to be seen as comprising a relatively minor subset of communication—of language conceived more broadly. What we have done in this book is, quite simply, to probe how the principles of communication that can be (and have been) identified in 'narrow' language are more common than has hitherto been appreciated in wider communicative encounter, and encounter beyond the human; and thus how they are constitutive not of humankind alone, but of naturekind. And as the sciences dismantle the basis for communicative discontinuity and hierarchy, the basis for wider forms of discontinuity, disconnection and claimed superiority is also fundamentally undermined.

The exploration that leads to these conclusions has involved probing findings that have emerged as a technological revolution precipitates a revolution in biology, and in particular what biological scientists increasingly reveal about capacities for communication in non-human animals, plants and other organisms, and the linguistic nature of this. This capacity for communication in some cases embraces dimensions once thought of as peculiarly human, including those of symbolism and culture. The ever accelerating discoveries that are producing these findings are often facilitated by the very technologies, from drones to AI, that in other contexts drive extractive and destructive relations with non-human life; yet they provide us with one basis for dismantling belief in communicative discontinuities within naturekind, and showing that its diverse denizens have more in common than we might once have imagined. Our argument has gone still further, however: challenging the more fundamental assumptions on the basis of which, in the history of ideas, boundaries between the human and the non-human have been constituted. This has involved theorising communication in a more 'symmetrical', or non-hierarchical, way that embraces all of naturekind. We have found inspiration in the emergent field of biosemiotics, given its attention to theorising the exchange of meaning—a concern that is lacking in much of the emerging biology of communication. The biological revolution has now rendered a theory of meaning as essential for biologists as it has always been to the understanding of human language, life and culture, in light of growing evidence of syntactical dimensions to communication, social learning and culture beyond the human. The field of biosemiotics has thus become essential to all biology and to the study of evolution, even if not all biologists yet realise this.

We have, however, found existing biosemiotics wanting, given the strong distinctions it makes between so-called prelinguistic signs and those with symbolic dimensions. Among the foundational assumptions of much existing biosemiotics is that there is something unique in human language, human social learning and culture. This idea, and the now superseded biology that once supported it, led biosemioticians to follow semiotic theorists who offered the promise of understanding how meaning could be conveyed in prelinguistic ways. We worry that even those seeking to redress the fallacy of a communicative separation between humans and the rest have been decoyed by this promise, adhering to the notion of signs as prelinguistic, and imagining that it is by focusing on signs so conceived that we would be focusing on a mode of communication that we share with naturekind. To attend to the linguistic mode, they suppose, would be to privilege what separates us.

It is this that we have rejected. Whilst the turn to biosemiotics is essential, we have returned also to the structural linguistic traditions of the human social sciences and humanities, releasing them from their anthropocentric roots and applying them to the understanding of communication beyond the human. We have thus pursued an approach grounded in a 'structural biosemiotics', to elucidate how meaning is created and shared within more-than-human worlds. We have demonstrated the potential for this through examples extending from companion relations involving chickens and horses, plants, bees and bats, through to communication beyond the human in whole assemblages, be they forests, soils, seas or cities.

Our considerations have ranged across parts of the pluriverse where the ontology of communicative capacities beyond the human, and the boundaries and hierarchies involved, are very different from those envisaged in scientific canons. Revealed, in many cases, are conceptualisations of communication amongst humans and with non-human natures indicative of companionship and assemblage relations that are less hierarchical, more mutual and connected. By taking a structural-biosemiotic approach we are thus able to appreciate more fully and respect anew ways of thinking and being that have too often been dismissed and subjugated in the sciences. As Amitav Ghosh has recently put it,

> What if it was the people who were regarded by elite Westerners as brutes and savages—the people who could see signs of vitality, life and meaning in beings of many other kinds—who were right all along? What if the idea that the Earth teems with other beings who act, communicate, tell stories and make meaning is taken seriously?[1]

Taking these ideas seriously is precisely what we have done in this book, in a move that is therefore also a challenge to the many injustices with which the sciences have coevolved. As we have also argued, however, the pluriverse here invoked is not just geographical, but extends across the multiple worlds—the multiple ontologies—that everyone occupies. As we hope to have demonstrated, linguistic communication beyond the human is a feature of everyone's everyday lives.

Our chapters have brought together ideas and experiences from contexts in Africa, Australasia, Europe and the Americas. In doing so we have combined material and reflection from our own experiences as anthropologists and dwellers in rural south-east England and West Africa, with the multispecies ethnographies and related works of many other scholars, practitioners and commentators. We have been deliberately selective, identifying examples where there are accounts rich enough to evidence a structural-biosemiotic framework, and deliberately reinterpretative, drawing out communicative dimensions sometimes left implicit in the original accounts.

All the examples show how communication is multisensory (multimodal), attuned to a being's sensory capacities and *umwelt* (its sensory and experiential world). The signing of chickens through clucking and flapping; of horses through touch, smell and sound; of plants through chemical and visual signals; of bats with each other and other animals through ultrasound; of soil organisms through chemicals and touch; of sea creatures through sonic-, visual- and movement-based signs attuned to their underwater worlds: these and other examples across the chapters have illustrated the basic tenet of biosemiotics, that multimodal semiotic processes operate across all life. They include, but go way beyond, the aural and visual signals of speech and writing on which the standard (narrow) conception of human language is predicated. We have explored how, through multisensory signing, beings communicate with others of their species (as chickens, urban pigeons, bats, bees or fish coordinate their paths and activities, for instance) as well as across species, often linking and holding together whole assemblages—whether the trees, plants and animals that constitute a forest, the mesh of animal, plant, fungal and bacterial life in a patch of soil, the assemblage of creatures and other life in a sea, or the mass of animal and plant life even in cities. What might have been considered mechanomorphically as forest, soil, marine or urban ecosystems held together by flows and feedbacks of nutrients, energy and instinctive behaviour are thus revealed as living, semiotically constituted, communicative communities, in which signing conveys meanings to which other organisms respond, setting off further processes of signification. In exploring seas and cities, in particular, we have considered further how water, wind, minerals and artefacts become part of such communicative assemblages, and in this we align both with vital materialist scholars who consider such entities as lively, and with framings in the pluriverse that treat them as alive: as manifestations of spirit, or similar. In the chapter on bats, we also touched on assemblages within bodies, in relation to the endobiosemiotic processes through which viruses communicate with cells.

Our focus has been on how meaning is created and shared in such communicative encounters, the conversations they become part of, and the communities so forged. Whilst we have probed this in some examples that do not involve humans (conversations amongst bees, bats and within soils, for instance), our central interest has been in communication across human–nonhuman divides: how people variously pick up on, read, interpret, join and become part of conversations and communicative communities beyond the human. We have probed this in more 'dyadic' companionship relations, where the examples of chickens and horses also show companionship to be a way of extending mutual capabilities. We have explored it also in situations where people interact with whole groups of a particular species, in the examples of bees in a hive and bats in a colony. And we have considered human participation in whole assemblages of forests, soils, seas and cities, in which communicative relations may involve multispecies semiosis within the assemblage, with the assemblage as a whole, or indeed with particular non-human companions who live there (as in the examples of human–pigeon and human–monkey communication in cities). As we have indicated, across all these, even though people lack much of the sensory or *umwelt* alignment to pick up on many of the particular signing processes of non-humans, meaning is still shared—conversations can nevertheless occur—through the structural linguistic processes that a structural-biosemiotic framework reveals and explores.

From the insights of structural linguistics and anthropology we can understand that, where signs are picked up, their meaning depends not just on the sign itself, but on its relations within wider fields of meaning: the presence (the not-absence) of a stick in human–chicken communication, and whether it is waving, or not; a droopy (so not an upright) stance that people register visually in a water-stressed plant; a scented darkness (so not a breezy light) conveying the meaning of being in a forest; grey hardness around and underfoot (not softness and greenness) conveying the meaning of being in a city, not countryside. Such relations are definitive of structural semiotics in the human sphere, and our examples suggest a centrality to communication beyond the human too.

Meaning is also created and shared through the ways that multiple signing processes combine in grammars, or syntax. Syntax determines the varied meanings recoverable from the combinatory possibilities of a given set of signs, but can also itself be an emergent outcome, as such combinations mesh with experience. We saw this in many examples, from the grammar of bee dances or pigeon navigation (with their sophisticated combinations of directional and visual signing) to how both humans and chickens combine calls, movement and touch syntactically when sharing meanings in conversations about 'getting in' (or not) to their roosting box at dusk. The grammars of human–horse communication in riding integrate touch and pressure with sound in varied combinations that gain meaning according to an established syntax. We saw how people participate in these signing processes and grammars, becoming part of conversations: whether involved in dyadic companionships in living and working with horses, bees or houseplants; engaging with the grammars of wider assemblages when cultivating crops in soils, immersing themselves in seas or forests, practising activities such as surfing, or simply in everyday life in cities.

Cases cited across the chapters above show how combinations of signs and grammars build up through experience, to acquire a 'packaged' quality. Thus our discussion of bees exemplified how it was through their experiences with a handler that signals emanating from them (of movement, bodily placing and touch, as well as smell) came to be meaningful to the bees, signifying 'not a threat', and how that packaged set of meanings in turn became a metasign—a habit through which these bees recognise their handler. Companionship relations with horses similarly reveal this kind of packaging, and the structural principles of human–horse communication through which signing is combined in grammars that become habitual over time. This has enabled interspecies languages to codevelop with human–horse interaction, with many regional and specialist dialects associated with different human–horse practices, from travel to racing, showjumping to ploughing. Packaging is apparent too in human–bat relations, even though in our West African examples these involve habitual relations of mutual coexistence more than engaged interaction.

Across the chapters, examples show how even where people cannot pick up on the details of the signs and grammars in play amongst non-humans, they might pick up on the higher-order outcomes of these, which become discernible as metasigns that acquire sense in relation to other metasigns. Thus in considering soils, particular fauna, colours and textures become metasigns in conveying meaning about soil quality. In considering forests, the presence of a patch of forest in an otherwise grassland landscape conveys a range of meanings about vegetation quality and the state of society locally. It also conveys a story about the human–non-human processes involved in creating the landscape, illustrating that metasigns can also be syntactical and take the form of narratives and genres (metasyntax) that convey more-than-human histories.

A structural-biosemiotic framework also captures how signs and grammars are emplaced, and how place affects meaning: how sea assemblages evoke different meanings in relation to particular shores and coastlines, for example. Spatial emplacement shapes how signs are combined in grammars, as pigeons combine their solar, magnetic and directional navigation signing in different ways according to the landscape around; or the grammars of horse riding may be inflected by the smooth surface of a dressage arena, or flinty hill paths. Spatial emplacement is manifest in human–chicken grammars as they emerge emplaced in one type of pen and door-opening technology, as compared with another in the neighbouring field.

Such emplacement is also temporal. Conversations are shaped by the rhythms of time of day, such that getting chickens into their house is a more straightforward conversation at dusk than it is earlier in the afternoon. Cultivation conversations with soils differ as the seasons progress through an annual cycle, and indeed, over years, decades and longer, as in the emergence of dark-earth soils. Conversations over such longer timescales also manifest in the entanglements of families with particular trees across generations, or in the entwining of human social identities with land and forests across centuries. The emplacement of trees (and human companionships with them) has a particularly enduring quality reflective of their stasis and longevity, which contrasts with the mobility and shorter lifespans of companion animals; so, as we have emphasised, it is not surprising that through conversations with trees, people often acquire their own emplacement and connection to place. Enduring buildings or artefacts in cities may have a similar character.

Emplacement shapes the meanings that derive from matter 'out of place', or matter or happenings 'out of time': disruptions to the normal more-than-human order of things. What is out of place can only be known in relation to subliminal grammars that constitute 'orders'. It shouts significance in ways that reveal those very orders—in usual circumstances unnoticed—that they have interrupted. We have seen this across many of our examples, from an out-of-place donkey stressing a horse, or a tractor panicking chickens, to plastic bags in seas, chainsaws in forests, out of the ordinary winds on a beach, or unseasonal heat. The extent to which matter is out of place or time surely depends on scale and point of view: it stands in relation to emergent orderlinesses that come to shape different conversational practices. As we have noted, people might in some contexts be out of place from the perspective of their non-human cohabitants: spooking animals in a field that have become accustomed to their absence, for instance.

Meaning, as these chapters have illustrated, thus depends not just on particular signs (and their qualities, symbolic or otherwise), but also on how they are combined in grammars or syntax; how experiences build up, over time, into packaged meanings—paradigmatic metasigns; how grammars or syntax themselves become enwrapped in higher order juxtapositions and narratives; how this packaging happens in the context of specific experiences; and how all this is emplaced, spatially and temporally. Some hitherto heterodox biosemiotics analysis has also drawn on these structural principles rooted in structuralist linguistics and semiotics (rather than invoking things prelinguistic), but our examples and analysis, drawing also on theorisations in structural anthropology, focus attention on the emergence of these higher order metasigns and narratives that 'package' information and meaning more efficiently, and on the orderliness that these convey, and thus the meaningful significance of disruptive matter out of place.

A focus on structure certainly does not mean that all is static. In our exploration of interspecies communication, whilst we have emphasised the significance of grammatical structures, we have also shown how these are forever emergent, lived and enacted in conversations. Indeed, the more stable dimensions of human language, as fixed in writing, for instance, might be considered as exceptions in this respect, for their relative stability. Our approach thus helps explain not just how communicative structures come to convey meaning, but also how they might change—as humans or non-humans shift their messages or communicative emphases, and repetition of shifted practices become embedded in new patterns, altered structures.

In all chapters, through a structural-biosemiotic lens, a different light has been cast on one of the key claims for the exceptionalism of human language: its supposed association with only-human capacities for intelligence and 'mind'. Our chapters have briefly touched on animal examples where 'human-like' cognitive capacities and intelligence are being revealed, such as amongst whales and dolphins. Yet in broadening our exploration of language through a structural lens, we have been able to displace the focus of this association, revealing broader language and inter- and multispecies communication amongst a wider range of beings, from bats, birds and bees to earthworms, plants and bacteria. We have addressed conundra as to whether collectivities (such as bee colonies) or 'super-organisms' (as cities have sometimes been understood) or ecosystems/assemblages (forest, sea, soil) can be conceived of as 'thinking', by showing how they become part of conversations in which meanings are exchanged.

Attention to structural principles does not deny agency or intentionality in communication, and the chapters above have explored many examples where conversations are initiated, led or redirected by beings in agentive ways. Non-human beings can certainly exert communicative agency, as abundant examples of horse, chicken and urban temple monkey behaviour have shown; but so can trees and plants, waves, or even a whole assemblage on land or in sea—inasmuch as a sign or metasign might convey a meaning to humans (or other beings), who then respond. Intention remains a more open question with regard to some of these beings and assemblages, but we have also explored how these questions of intentionality are resolved in very different ways across the pluriverse. Framings and ontologies in which trees have personhood, waves are the expression of a wind deity or the embodiment of a serpentine water being, or land and forest forms articulate their entangled co-becoming with particular groups of people, allow possibilities of intentional communication in ways overly foreclosed by the framings of current science.

Our structural-biosemiotic framework also challenges assumptions that communicative capacities and patterns are species characteristics—that tigers speak Tiger and whales Whale. Clearly there are aspects of semiosis that reflect any being's biological make-up, sensory capacities and needs, and these, as we have suggested, might be treated as biological anchors. These anchors might be shared across all of a given species or subspecies, breed or type, as we noted in the cases of horses, bees, bats and trees. Yet within a structural-biosemiotic framing, such anchors can be seen to combine with other signing processes as part of grammars and emplaced packages, in configurations linked to experience. Species-wide generalities become less determinant. More relevant is the particular horse, or group of bees or bats, or tree: their particular historical experiences and what meanings they are sharing with particular other beings (people, perhaps) in particular contexts.

The communicative capacities and patterns that we have been documenting are not just a matter for individuals, of whatever species. Communicative encounters and conversations are relational, and shape each party, in a process of co-becoming that changes both. Just as Donna Haraway said of human–animal companionships, that 'partners do not precede their relating', so our examples show that participants in conversations beyond the human do not necessarily 'precede their communication', inasmuch as co-becoming can include changes to mental states and physical and bodily form: as, for example, in the mutual personality development and muscling-up of horses and riders through their conversations in dressage, or the very different example of viruses in human bodies provoking shifts in attitudes and health. Extending the principle to wider assemblages, we have noted too how a particular forest patch or soil area co-becomes—acquires its form and character—through its communicative entanglements with humans, as well as other beings.

Analysing communication through a structural-biosemiotic framework therefore affords no fundamental grounds for human exceptionalism and hierarchy (beyond the fact that all species are exceptional in their own ways). At core, meaning is and can be shared in similar ways amongst all beings and with their surroundings. In turn, where communication beyond the human is indeed attuned to and mobilised to serve human domination and superiority, this needs to be analysed for what it is: as part of, and supporting, relations of power.

Our examples have not shirked attention to power. They have acknowledged that companionships often take place on terms set by the humans involved. They acknowledge the power inherent in long-term processes of domestication. They have tracked many situations in which animals, plants and ecological assemblages are being applied to human-defined and prioritised social and economic purposes, whether in cultivating crops (see the chapters on soils and on forests, for instance); in keeping or hunting animals for food (chickens, bats, fish); in extending capabilities in work (in the cases of horses used in transport, bats and the military, dolphins in fishing); in protecting health (in discussions of bats and viruses, and of urban mental health); and in sport and recreation (in riding, surfing, chicken- and beekeeping, for instance). Yet each chapter has also told a story whereby fulfilment of these human-defined purposes depends on, or at least is better achieved by, aligning with non-human natures, and depends too on communication in support of this: in better discerning the 'purposes' and meanings of non-human beings and assemblages, and then attuning conversations to respect and sustain these. Power asymmetries are still present, but to degrees and in ways very different from when humans claim exceptionalism and top status in a hierarchy of beings on the basis of a narrow version of language, dismissing and denigrating other beings and assemblages as mechanical or only mechanistically communicative.

The stories told in our chapters do not just track how communication is part of physical and practical entanglements across naturekind. The connections established and sustained also implicate other aspects of sociality, extending to those of identity, community and emotion. Language is important to affiliation and community amongst people, and common identity is often found in the languages and accents that unite and divide us. Speaking multiple languages is not just practically helpful, but helps transcend narrow ethno-linguistic affinities in a cosmopolitan world. Capacities to communicate cannot be divorced from questions of affection and from the feelings of mutual understanding and trust that help humans live together, as social beings. Appreciating the linguistic basis of communication beyond the human, then, enables us to understand how this realm of sociability and affect widens, even if it is just a matter of a grunt of mutual acknowledgement in exchanges with a tiger that enables cohabitation, or—as in examples we have considered here—the sense in a forest that one is amongst friendly kin, or the regular movements of bats that signal their peaceful membership of a village community.

Crucially, the character of communicative encounters is not independent of the way people have come to reflect on them and on the possibilities they represent. We communicate differently with a chicken depending upon whether we understand its responses as instinctual or as communication that enables intersubjective appreciation. We engage differently with a tree if we understand it as embodying living forces or as the home of bush spirits, than if we look upon it as 'just' wood, bark, twigs and leaves, or indeed do not even see the tree for the forest. How and how easily we communicate beyond ourselves—and how aware we are of doing so, and of the grammars/syntax and the field of (packaged) signs in play—thus depends not only on experience, but on existing grammars and lexicons, and higher order packaged narratives and cultural grammars and lexicons that frame the world. We do not learn a new human language independently of our existing ones, and the languages that we develop beyond the human are no different in that respect: they are not independent of extant narratives; of existing packaged syntax.

Viewed thus, our relations with companion animals and working animals, even with insects, trees and soils, blur into our relations with each other. Just as families bond around the pets that give routines to everyday life and allow projections of warmth and intimacy, so people, as we have seen, might bond around a tree, or the performance of a task with an animal or patch of soil— whether of an ancestral settlement or a contemporary allotment. There is no need to anthropomorphise to acknowledge intimacy: a horse or a chicken does not lose its significant horseness or chickenness as it becomes attached to its human companions. Nor does one need to know the 'mind' of a companion, just as one cannot fully know what is in the mind of a human friend, lover, family member or colleague. Such companionship has to be worked at, just as it does with people, but this is not because animal or plant companions are 'stand-ins' for humans; it reflects rather a genuine cross-species ability to live meaningfully together. Exploring the communicative dimensions—how intimacies are built through the sharing of meaning, and how this builds up through experience—thus adds to works arguing that 'making kin' is not something confined only to humans, or human substitutes.[2] Companionship between people and other animals may be easily recognisable, indeed obvious, in the pet-keeping practices of late industrial societies, where 'companion animals can be horses, dogs, cats or a range of other beings willing to make the leap to the biosociality of service dogs, family members, or team members in cross-species sports'.[3] Companion relations extend much more widely, however: around the world we see a huge diversity of joint lives bonded in 'significant otherness', with creatures ranging from mammals to birds, reptiles, fish, insects, fungi and more, as well as with plants and the places they inhabit. Communication—conversation—is vital to these intimate relations; but those who have been taught to consider communication as narrowly linguistic can be blinkered in this regard, presuming that animals (let alone plants) lack what it takes.

Engaging with animals (and other non-human beings) can, moreover, be important to one's own understanding of who one is. As Dominique Lestel and Hollis Taylor write,

> Humans form their self-representations not *in opposition* to animals, as all Western histories of human evolution recount, but *with* them and *through* them. In other words, to be human does not mean to have fled animality, but on the contrary to live within it and to let it live within us [...]. A shared life implies becoming oneself while being other.

They note too that what the significant animal (or plant, or river, or spirit-being) is can vary from person to person, context to context:

> Animals that are significant to me are not necessarily so for my neighbour, even if we have 'animals in common'. What is important is that humans always form their self-representations by means of other animals, not the fact that they do it locally with this or that animal.[4]

Communicative companionship relations are not constant. Many are fleeting, or have fleeting moments within them of devolved self, of intensely integrated, communicative experience. But these may be interwoven with much more mundane relations in which one coinhabits a place or exists in an ongoing relationship with a non-human. This is how many of us engage with our pets in the course of a day, only punctuated by occasional moments of intense engagement. It is what our chapters above track in relation to companionship with chickens, horses, bees and bats—and with the last of these in particular, it is simply patient cohabitation that predominates. Likewise, in an established human partnership, a friendship or marriage there is much that is mundane and routine; but there can be fleeting moments of a different kind, moments of deep emotional or mental connection through a certain conversation. We know this in regard to our personal human relationships, including sexual encounter, and we also know that those moments, however fleeting, are important in sustaining the whole relationship. Many such moments of intimacy are particularly 'animalian', in our oscillation from our social self to our animal self, stroking and grooming our partner and switching into the sexual self, casting aside the clothing of our social personas to expose what we are apt to see as our animal selves, sensing our 'animal side'. Understanding multispecies relations becomes important to understanding our own animality, not just our multispecies sociality.

How does one square such worlds of intimate connections and companionship with the fact and practice of violence towards companions: animals incarcerated, bred, torn from their parents, traded, castrated, spayed; and with killing—of animals and of plants? A common resolution has been to eliminate the possibility for such intimacy: through detachment, through separation and the denial of connection, through extension of the dehumanisation that enables killing in war or genocide, and through what Achille Mbembe calls 'necropolitics'.[5] This is to render non-human animals, as well as people, the 'living dead', withholding from them liberty or autonomy over their lives, in the same way as necropolitical regimes enslave, imprison, segregate selected human others and expose them to mortal danger and death in zones of existence where people no longer have sovereignty over their own bodies.

Such detachment and separation, such denial of intimacy, is hard to achieve in practice, and is usually incomplete. Abattoir workers often experience trauma,[6] and even a hardened farmer finds it hard to euthanise an animal. There is crossover between animal and human killing fields, in the form of the 'modern, state-employed mass murderers' who make murder, in Hannah Arendt's terms, bureaucratic, and systematic.[7] There are, however, other ways to respond to the coexistence of violence and positive affect in our relations with animals, and one such is to kill with respect. Pet owners feel anguish and grief when they must euthanise companion animals, and the act is carried out with respect—borne of intimacy, not separate from it—with many rituals concerning burial, the spreading of ashes and memorialisation echoing those adopted for fellow humans. Some even find it respectful to consume the chicken they have killed after it has become too old to lay eggs.

Around the pluriverse, where intimate, communicative interspecies companionship is often taken for granted, we find a similar crossover, navigated in similar ways. Among some peoples of the Philippines, as we have seen, deferring to the god of bees is necessary for permission to destroy a hive. Many North American hunting peoples conceive of animals as other-than-human persons who give themselves up to hunters; their gift incurs 'a debt that must be re-paid through the performance of certain ritual practices', and hunting becomes a 'long-term relationship of reciprocal exchange between animals and the humans who hunt them'.[8] In parts of West Africa, only certain hunters kill: those renowned for their skill and abilities to 'know the bush' and 'know the animals', and to communicate with them in a way that ordinary people cannot.[9]

Multispecies companionship might also be cited in support of arguments against the slaughter of animals. Here we are close to the Pythagorean insight, that abstaining from animal food is productive of peace, 'for those who are accustomed to abominate the slaughter of animals as iniquitous [...] will think it to be much more unlawful to kill a man, or engage in war'.[10] Pythagoras avoided killing or injuring animated beings for food or sacrifice, and instead observed 'most solicitous justice towards them'. Political leaders especially, he considered, should avoid eating animals:

> For as they wished to act in the highest degree justly, it is certainly necessary that they should not injure any kindred animal. Since, how could they persuade others to act justly, if they themselves were detected in indulging an insatiable avidity by partaking of animals that are allied to us? For through the communion of life and the same elements, and the mixture subsisting from these, they are as it were conjoined to us by a fraternal alliance.[11]

This perspective emerged within the prevailing ontology of Pythagoras's world, which allowed for people being reincarnated as animals—a framing still shared by some religions. One does not need to subscribe to a particular global religion, however, to appreciate multispecies companionship. Interspecies empathy, and the communicative connectivities it is associated with, can lead to a decision not to kill, or to do it differently and respectfully.

Violence can cut both ways. The sense that surfers feel, that the power of the waves could flip and drown them, is ever-present to them—and indeed is part of the draw, echoed in many extreme sports. The sudden switch to danger when exuberant bullocks do not disperse, but approach in hostility, surprises walkers on the English South Downs, just as the comfort and security of a tourist car in an African national park evaporates when the vehicle angers a bull elephant. To the extent that the desire to avoid such confrontations is rooted in constant awareness of the power of the living beings and other elements with which we cohabit to 'strike back', this surely derives less from explicit knowledge of their characteristics and behaviour than from knowing the codes in an interchange of meanings: the language beyond the human that can help relations remain peaceful and respectful. We authors saw this constantly with our Guinean tutors as we immersed ourselves in their forest worlds: how and where to step and where not to, which streams to take water from and which to avoid, which plants to cut and which to keep—these were facets of a conversation in which human desires and intentions were in constant interplay with a world cohabited by animals, plants and land-spirits whose interactions one engaged with, but was only a small part of. Such a process of learning respect and the limits to one's control is also a kind of 'becoming with', constituting our humankind which is also naturekind.

Political Naturekind

As we have been arguing, and we hope demonstrating, all people are inevitably entangled in wider communicative relations within naturekind, however much we might adhere to notions of human exceptionalism, construe ourselves as separate and unique, and treat nature as effectively 'zombified'. Kath Weston, Anna Tsing and others have been drawing attention to the new intimacies that people inevitably establish within an ecologically damaged world, albeit without probing the communicative dimensions of these entanglements.[12] It is important also to acknowledge ruptures of communicative intimacies, however, given the collapse of life around us. There is a need to recognise that changing social and livelihood-related routines that intensify human–human communication can draw lives and attention away from routinised encounter beyond the human. And it is valuable to reflect again on separations between human and non-human natures, their effects in rupturing communicative connections, and the implications of all this. These separations are embedded in wider politico-economic and discursive relations that contribute to the destruction of nature—and also, paradoxically, to many of the approaches to conservation that purport to save it.

Thus we can reflect on the separation between humanised places and the 'spaces for nature' sometimes associated with biodiversity protection and restoration schemes, whether in national nature reserves, wildlife parks and forests, privatised rewilding projects, or even the 'sacrifice zones' altered and then abandoned by corporate business. Spatial separations have been given new impetus by ambitious global targets to conserve 30 per cent of land- and seascapes by 2030, whether through strictly protected areas or potentially more people-inclusive 'Other Effective Area-based Conservation Measures' (OECMs). Such spatial separations potentially inscribe more fundamental divisions between 'nature' and 'culture' into experience of landscape itself. The ruptures created by forced separations are not just physical, but communicative, splitting people from their more connected experiences with more-than-human worlds. Those displaced on the frontiers of settler colonialism, those disqualified as 'nimbys' on the frontiers of development, those dispossessed on the frontiers of 'green grabbing' (the appropriation of nature for environmental ends),[13] and others so displaced, experience such intense communicative rupture that it inevitably provokes an emotive response: resistance.

We can consider, too, how spatial divisions interact with market-based approaches to conservation: approaches that strip or divide assemblages into 'bits', re-envisaging components as units of biodiversity, tons of carbon, cubic metres of water aquifer, doses of nature for green prescriptions and so on. Such splitting into measurable, bankable units and commodities may help bring business and finance to conservation through credit and offsetting markets, but it also presupposes deeper conceptual separations between people and nature, and assumes their possibility. Commodification and offsetting processes for biodiversity are premised upon substitutability—that one unit is equivalent to another, one exemplar of a species is replaceable by another; a substitutability that is convenient for late industrial capitalist development and the powerful new narratives of so-called 'nature positive' business, but that is revealed as illusory in light of the importance of place and of culture in the lives of all beings. Indeed, the latter suggests that discussions of biodiversity should no longer be confined to species or genetic diversity, or diversity of habitats; they must also comprehend the more-than-human cultural diversity codeveloped through social learning in particular places.

We can perceive how all these separations help configure (and are configured by) the discursive shaping of 'nature' in science, policy, education and advertising that enables us to suppose that it is fundamentally different from human life: whether as a machine-like ecosystem, empty wild landscape or green space, or a set of resources or commodities that can be used instrumentally. Given that contemporary politico-economic and legal systems are predicated on such reasoning, the revolution in biology that gives the lie to it and on which this book is built has implications that are truly revolutionary, in the political sense of the term. The idea of language, social learning and culture as uniquely human has become part and parcel of these deeper physical and conceptual separations that are in contradiction with emergent science, so we must inevitably reflect on a 'political naturekind'.

Even though human–nature separations are rooted in and reinforced by politico-economic and discursive processes, such separations are not inevitable. We can also discern pathways through which human and non-human natures might thrive together, premised on what unites, not what divides.

Appreciating these, and making space for them, is integral to political naturekind. In addressing this issue, we are adding to the many works in the ecohumanities and post-human social sciences which now argue that framings beyond the human are not just necessary in social analysis, but have a wider intent, in terms of finding better modes of living together. We authors of this book join academics, activists and Indigenous peoples alike in emphasising the significance of connectedness beyond the human in environmental politics, and the overlapping narratives they have developed variously emphasising conviviality, recoverability, care, intimacy and justice. But what can our systematic focus on communicative dimensions add to these?

Arguments have been advanced in favour of recognising and reviving ways of being that are not premised on human–nature separation, but on living together in what Emma Marris terms a 'rambunctious garden'.[14] For example, Bram Buscher and Rob Fletcher argue for more 'convivial' modes of conservation,[15] inspired in part by the ways of living of those marginalised from mainstream modernity, including Indigenous people and pastoralist communities. Indigenous scholar Joy Todd argues that such peoples should be credited as the originators of these arguments, which challenge colonially derived ways of thinking and acting.[16] Calls to respect and build on Indigenous philosophies, and to challenge the hierarchies that subordinate nature to a particular version of modernist culture, are widespread, from local social movements to the debating halls of global biodiversity negotiations. Building on longer traditions of 'community' conservation and an array of 'co-governance' and 'community engagement' approaches,[17] it is now commonplace for external conservation and restoration policies to advocate the safeguarding of peoples' human rights and rights to places, and recognition of and support for Indigenous traditional territories within area-based conservation approaches. For all the good intentions, however, much policy and action remains tokenistic, buying local support for protection of a separate nature with 'alternative livelihoods' and the sharing of the financial benefits from capitalist conservation schemes. At their best such initiatives recognise local ways of being and support both the human and non-human worlds; at at their worst, however, they police Indigenous people into 'staying Indigenous' by one criterion or another: getting them to 'perform Indigeneity', and commodifying heritage alongside conservation. This can produce incongruities that are often ignored: Indigenous people who themselves seek to acquire education or private property risk 'de-authentication' and privation of the benefits of recognition. It maintains a separation, not between all humans and all nature, but between some humans—those seen as destroying nature—and those seen as living well with nature (being 'part of nature'). The characterisation of Indigenous people as 'part of nature' has for a century been construed as an abuse: a form of othering that suggested that they were somehow not part of civilised, contemporary humanity; but its current inversion maintains the separation—as if being part of naturekind were not in fact common to us all. Erin Fitz-Henry argues that many such accounts, 'despite their insistence on ontological multiplicity, end up over-playing and thereby unduly solidifying ontological contrasts between the "West and the rest" in the service of producing a much-sought-after antidote to Western naturalism and liberalism'.[18]

One way out of this bind is to be inspired by a call to respect the pluriverse; to recognise and learn from Indigenous worlds, but to realise that the communicative roots of conviviality are not exclusive to those with particular beliefs and historically interconnected relationships with non-human natures. A focus on communication across naturekind reveals the existence of, and potential for, convivial relationships amongst us all, in everyone's everyday lives, everywhere. We have argued here that communicating in naturekind— building languages with the beings and worlds around us—is an inevitable facet of being alive. We have tried to show this by deploying just a few examples of something that is in fact all-pervasive: discerning it in the ways people relate to their companion animals, to plants, and assemblages. We therefore do not need to look to the marginalised peripheries of the world, but can look to ourselves, wherever and whoever we are.

To stay with European contexts, then: Steve Hinchliffe and Sarah Whatmore, for instance, argue in support of approaches to more-than-human conviviality in 'living cities' in the UK, where 'urban liveability involves civic associations and attachments forged in and through more-than-human relations'[19]—the communicative dimensions of which we too have been excavating. More broadly, a new narrative of a more 'recoverable earth' has been discerned by Paul Jepson, led by stories of rewilding and restoration that 'tell of feelings of despondency and processes of awakening, action, and reassessment leading to the recovery of natural and social well-being'.[20] Jepson traces the origin of this disposition towards the world around us to the narrative structures of accounts of recovery from mental health challenges; it is certainly one that emboldens grassroots practitioners and activists to assert their ability to lead change locally and produce better outcomes for human and non-human sociality alike. National-level support for such local initiatives is also evident: increasing nature-connectedness has become a core focus within the UK government's Nature Recovery policies, for instance, encouraging educational and urban design interventions to increase sensory, meaningful and emotional engagements.[21] Again, by excavating the communicative dimensions involved, we align with all this, but amplify it, in the conviction that conservation, restoration and recovery are not just about human safeguarding of a separate nature that provides our 'life support'; not just about the new commodification of this life in the optimistic assumption that we can 'sell nature to save it';[22] but about recovering the intersubjective communication that is life.

An emphasis on care, long central to feminist analysis and advocacy, has also become central to the envisioning of ecological relations and futures. Joan Tronto has outlined how care can be taken to encompass 'everything that we do to maintain, continue and repair "our world" so that we can live in it as well as possible. That world includes our bodies, our selves, and our environment, all of which we seek to interweave in a complex, life-sustaining web.'[23] As Maria Puig de la Bellacasa argues, matters of care are about care for and with a more-than-human world.[24] Fostering and enabling caring relations between people and other beings becomes central to envisaging and enacting transformations to more sustainable ways of life.[25] Eileen Crist discerns how such care is deeply entangled with intimacy, and is central to an 'ecological civilisation' that eschews illusions of human supremacy and pursues freedom for all, not just human, life.[26] And as we have been arguing, by examining communicative dimensions, we reveal the forces of affinity and kinship with the world around.

Expanding notions of conviviality, recoverability and care to naturekind as a whole shifts the emphasis of environmental politics to recognise and respect more-than-human collectivities and freedoms. In pursuing communicative dimensions that are relational and intersubjective, we encounter relations with a world that cannot be 'owned' in quite the ways that a separate 'nature' can be. A multispecies commons is revealed: a communicative commoning of the kind driving many decolonial struggles, Indigenous people's struggles, and indeed ecofeminist struggles, when the everyday labour routines of disempowered, disenfranchised and dispossessed women are entangled with more-than-human ecological concerns.[27]

Our focus on communication has important implications, moreover, for the emerging conceptualisation of multispecies justice, earth law and the rights of nature.[28] This broadens theories of environmental justice from a focus on the rights and justice claims of people, to comprehend justice beyond the human and for interconnected assemblages of the human and the non-human.[29] The 'rights of nature' movement extends legal standing and the capacity to litigate to particular animals, trees and rivers, and more, and some countries recognise such rights in national constitutions. The government of New Zealand has declared the Whanganui River and Mount Taranaki to be living entities holding legal rights equivalent to those of humans, and in India the high court of Uttarakhand ruled that the Ganges had similar rights (although this is contested at state and federal level)[30]—cases that build on the earlier decisions to grant rights to 'Pachamama' (World Mother) in the Bolivian and Ecuadorian constitutions. Such approaches eschew assumptions that only humans can be treated as moral or rights-bearing claimants, opening up forms of recognition and respect that have long been present in parts of our multi-world pluriverse, but need no longer be exclusive to these. Celermajer and colleagues note that '[a]lthough we cannot speak for all Indigenous approaches [... themes] of reciprocity, respect, sustainability, and spiritual–material interconnection echo throughout Indigenous philosophies across the globe'.[31]

An appreciation of a communicative, intercultural naturekind prompts new ways to reflect on the apparent trade-offs between 'rights of nature' and 'rights of people'; between non-human and human justice,[32] as we appreciate just how inextricable these are: how rights of nature may be essential for people to be able realise their right to a healthy and clean environment, as enshrined in the UN Declaration of Human Rights, for instance. It reframes how to consider relations between questions of human discrimination and discrimination beyond the human, extending the long-standing concerns of critical race and class theorists and feminist scholars to debates about multispecies justice—potentially, it is to be hoped, in the cause of a careful building of solidarities across these struggles, without collapsing their distinct concerns.[33] The mercantile forces reshaping our planet in the Anthropocene are not just a human thing. As has long been argued (but long ignored), there is no economy outside 'the œconomy of nature',[34] and inscribing a communicative dimension into the way economies are construed must inevitably inflect the current orthodoxies aligning Darwinian evolution theory and economic theory. Viewing matters through a 'more-than-human' lens serves to clarify not that the current economic order has produced problems for nature and for some people, but that capitalism in all its forms (as too the socialistic counter-creeds it inspires) is an inherently multispecies phenomenon. Our era must contend with multispecies capitalism, out of which emerge multispecies contradictions and struggles. Moreover what is united in such struggles is not just collections of species, but collections of emplaced cultures that have codeveloped common languages.

There is an irony in the fact that the emergence of our current economic order has depended on human–non-human communication, co-optation and coercion (in incarceration, training and domestication, industrialised production and consumption that destroys ecologies), whilst at the same time denying this, through the mechanomorphism that humans revert to in justifying their actions. Equally ironically, the technologies that are enabling enhanced human–non-human communication (sometimes designated 'Nature 2.0') are also enabling increasingly efficient extraction of the labour of life beyond the human.[35] Attention paid to communication, however, allows alternatives to come into focus because, as we have seen, not all collaborations between humans and non-human natures involve exploitation without a respect that is rooted in communicative encounter. Such attention offers the potential for building new solidarities and resistance. This more hopeful emphasis is the theme of our last section, looking at how structural biosemiotics can help foster a more positive politics for naturekind, in relation not just to the goals of environmental politics, but also to the actual 'doing' of the politics.

Doing Political Naturekind

Prevailing mechanomorphic perspectives on nature in the sciences are functional to politico-economic forces. Effective challenging of the modes of separation will not be achieved simply by the promotion of 'new ideas' of the sort this book has developed, as such separations are inscribed in the world and its infrastructures as much as in our minds: in the ordering of landscapes, in food systems, in planning legislation and legal precedent, and so on. Fostering more convivial, caring, just relations can nevertheless begin with new understandings, as the sciences come to produce findings that require attention be paid to the making of meaning across naturekind, even if new understandings come up against old infrastructure and practices that have become entrenched over the past five hundred or more years, and on which current human lives depend. What kinds of political processes can foster the constructive disruption necessary?

To address this, it becomes important to integrate the ideas we have been developing here to a degree with wider questions regarding 'sustainability transformations': those 'fundamental changes in structural, functional, relational, and cognitive aspects of socio-technical-ecological systems that lead to new [...] outcomes'.[36] Some have been finding answers in the need for fundamental structural changes to production and consumption. Others favour speeding up more incremental transitions, through combinations of technological innovation, economic incentives and progressive policy. Others again have been suggesting that change will emerge from below, through networks of civic movements and grassroots activism that, together, construct wider change.[37] There is a vast array of work on ecological-political transformations,[38] and reviewing it is not the task of this book. So how can these debates be enriched by a work whose aim is to discern the lineaments of communication and the making of meaning across naturekind, including in politics?

The concerns of such a political naturekind encompass both political goals and objects, and political processes: that is, how change happens. Yet political theory, like social theory, has focused to date on humankind, not naturekind. In mainstream canons, political actors, political subjects and political processes are all human, albeit interdependent with 'resources'; the non-human is neglected, or actively excluded.

Karl Marx did theorise the rupture between humanity and the rest of nature emanating from capitalist agricultural production and the growing division between town and country, but he phrased this in terms of a 'metabolic' rift, not a communicative one. He contrasted the free, conscious productive activity of human beings with the unconscious, compulsive production of animals. He conceived of the free, universal, creative and self-creative activity through which people create and change their historical world and themselves (which he called 'praxis') as something unique to humans, in this distinguishing them from all other beings.[39] It was a nature of the living dead that Marx left space for when contemplating the key historical moments at which contradictions or mismatches emerge in relations of labour and capital. Others in the Marxist tradition, such as Lenin and Gramsci, dropped nature entirely, focusing on the class tensions wrought by processes of differentiation in society—human society, that is.[40] When Karl Polanyi drew attention to how crises and tensions in the relationships between economies and societies shift, and how these conjunctures generate transformations, it was of human societies that he wrote.[41] And when Hannah Arendt criticised political philosophy for its over-emphasis on the contemplative life (*vita contemplativa*), neglecting the active life (*vita activa*),[42] she emphasised the importance of everyday political action or praxis, viewing action as a mode of human togetherness. She developed a conception of participatory democracy which contrasts in its liveliness, inclusion and radicalism with more bureaucratised and elitist forms of politics; yet the togetherness emphasised—the connectivity—is envisaged as solely human.

Reflections on forms of political action, agency and activism have variously been formulated around class, race, gender, place and their intersections, and in considerations of social movements and environmental social movements. Questions of identity are central, and so, as Nancy Fraser argues, politics and justice are about recognition and respect as much as about procedure and redistribution[43]—yet such recognition and social identity is generally restricted to people.[44] Theories of contentious politics have explored the capacity of collective mobilisations to reframe agendas, expose hidden power, challenge dominant interests and bring alternatives to the fore, whether around economic class interests or emergent social or environmental issues,[45] but the collectives envisaged in the doing of contentious politics are envisaged as only human. But do those involved in all this not find inspiration in contemplative walks or swims? Did Karl Marx not play with his dogs? Novelist Marian Comyn, a close friend of Marx's daughter Eleanor, described how 'Karl Marx was fond of dogs, and three small animals of no particular breed—of a mixture of many breeds indeed—formed important members of the household'. She goes on,

> One was called Toddy, another Whisky—the name of the third I forget, but I fancy that, too, was alcoholic. They were all three sociable little beasts, ever ready for a romp, and very affectionate. One day, after an absence of six

weeks in Scotland, I went to see [Marx's daughter] Eleanor and found her with her father in the drawing-room, playing with Whisky. Whisky at once transferred his attentions to me, greeting me with ebullient friendliness, but almost immediately he ran to the door and whined to have it opened for him. Eleanor said: 'He has gone down to Toddy, who has just presented him with some puppies.' She had hardly finished speaking before there was a scratching and scrambling in tile hall, and in bounded Whisky, shepherding Toddy. The little mother made straight for me, exchanged affabities in friendly fashion, then hurried back to her family. Whisky meanwhile stood on the rug, wagging a proudly contented tail, and looking from one to the other, as who should say: 'See how well I know how to do the right thing.' Dr. Marx was much impressed by this exhibition of canine intelligence. He observed that it was clear the dog had gone downstairs to tell his little mate an old friend had arrived, and it was her bounden duty to come and pay her respects without delay. Toddy, like an exemplary wife, had torn herself away from her squealing babies, in order to do his bidding.[46]

Cannot such appreciation as this be drawn into, rather than excised from, political reflection?

Political traditions of deliberative democracy have communication and language at their heart. Yet these, too, are human-centric, from their roots in Aristotle's notion of politics to its influential contemporary expression in the work of Jürgen Habermas and beyond. For Habermas, politics allows people to organise their lives together and decide what common rules they will live by, which depends on political argumentation and justification—processes of deliberation—and much writing on contemporary environmental politics is rooted in theories of deliberative democracy. Yet even when deliberating about ecologies, participants are assumed only to be human. And the kinds of communication they engage in is usually assumed to be of a narrowly human kind, grounded in formal human speech and writing.[47]

The ways most contemporary scholars lead us to think of political processes, even in engagement with environmental politics, are focused not simply human-centrically, but on human exceptionality. Political agency is confined to humans. If non-human beings—animals, plants, microorganisms, the wider assemblages they are part of—figure in mainstream political thought and approaches, it is as passive objects, for and on behalf of which people speak, not as active agents. Can this be changed? Can non-humans become participants in these processes? Can Karl Marx's Whisky finally gain recognition? Can political processes take place within and be shaped by naturekind? There have in fact been some moves towards multispecies deliberation and communication. What might be the insights and potential contributions from a structural-biosemiotic paradigm and framework, such as we have charted in this book?

Concepts once developed to guide and understand relations between people, such as citizenship, sovereignty, cosmopolitanism and republicanism, are now being reworked by some to shed light on the various ways in which other animals relate to human political communities.[48] Citizenship theory is being extended to include companion animals and farm animals as citizens.[49] International relations theory is being taken beyond the human, suggesting that international dimensions of states—sovereignty, territory, security, rights—are better understood as forms of interspecies assemblage that both generate new forms of multispecies inclusion and structure forms of violence and hierarchy against human and non-human alike.[50] All this shows naturekind to be part of political processes in 'interactional' ways. Bruno Latour saw political action and agency as dependent on actor-networks, comprised of both actors and actants, intentional and unintentional, living and non-living. It is the network as a whole that has power and the capacity to effect and shape political change.[51] In a slightly different way, philosopher Jane Bennett extended her conceptualisation of vital materialism to politics by arguing that objects possess power and agency by dint of being located in assemblages of human and non-human bodies, and living and non-living elements. In all of these approaches, however, there is little attention paid to the communication at work: to how communication beyond the human constructs and holds together political communities, let alone to the structures through which this happens.

The door is left a little wider for such communication in the 'cosmopolitics' of Isabelle Stengers, for whom political knowledge and positions are 'co-fabrications' in which all those (humans and non-humans) enjoined in it can, and do, affect each other.[52] Political events involve 'the management, diplomacy, combination, and negotiation of human and non-human agencies'.[53] Here we begin to see greater appreciation and attentiveness to communication beyond the human in political deliberation. A second entry point can be discerned in Hinchliffe and Whatmore's 'politics of conviviality', in the context of which they understand the collectivities and associations involved to be interspecies. They extend analyses rooted in feminist emancipatory traditions that replace the disembodied, rationalist ontological precepts of liberal individualism with a notion of political subjectivity as embodied. This removes the tie between political engagement and human language, allowing for recognition of engagement by the more-than-human, but through things 'prelinguistic'. As they argue, '[i]nstead of placing the political in the realm of conscious judgement and knowledge, here politics is extended to the "hinterworld of affectivity"—where intercorporeality exceeds the consciousness of "I think" and the "said" of language'.[54] Such works refer little, if at all, to biosemiotics, yet stand to benefit from its insights—and from debates over the sharp linguistic–prelinguistic divide (a framing that we ourselves reject). Acts of choosing, including choices and claims that might be termed political, are inseparable from acts of meaning-making and semiosis, which extend across all life-forms.

Some recent works concerned with deliberative democracy also welcome life beyond the human. John Dryzek and Jon Pickering admit non-human participants into their 'politics of the Anthropocene', encouraging 'a deliberative understanding of democracy—with meaningful communication at its heart'. This, they argue, will help render democratic institutions more responsive to signals from the natural world, and so, reflexively, modify their values and practices accordingly.[55] Dryzek and Pickering's perspectives are also inspired by examples from advances in biology that reveal sophisticated, meaningful communication amongst non-human animals and plants, but exactly what this communication consists of is less explored, attention being paid to signals, but without elaboration. They stop short of considering non-humans as able to exercise what they call 'formative agency' in such deliberation, considering that 'it is likely that only humans have the capacity to engage directly in the abstract reasoning and sophisticated communication required to shape what core societal values and concepts should mean'.[56] Yet they allow that people and their institutions can exercise formative agency in deliberation on behalf of non-human nature, or that they may do so through a productive partnership, both of which principles are implied in readings of examples such as the court decisions in New Zealand and India, referred to above, with regard to the rights of mountains and rivers. In order to be able to act in this way, Dryzek and Pickering suggest, people and their institutions need to become better listeners and interpreters of signals (or indeed the screams of pain) of non-human nature.

Some consider how animals, plants, ecosystems, regardless of agency, might be represented or 'spoken for' in political fora by human interlocutors or ombudspeople. Traditions and experiments such as 'Councils of All Beings' around policy issues are now in play around the world.[57] Yet such approaches beg huge questions about which people speak for which non-human beings, and what meanings they convey. A farmer used to the signs of living soils and their histories would 'speak for' them in ways very different from a soil scientist. A person might be assigned to 'speak for the bats', or a particular bat species, in such a council; yet which actual bats, where, would they speak for, in worlds in which, as we have seen, meanings depend on their particularity and emplacement? A structural-biosemiotic approach to language and culture beyond the human has the potential to refine such policy experiments.

Philosopher Eva Meijer goes one step further, arguing for a reconceptualisation of political deliberation, activism and democracy as an interspecies matter in which non-humans are indeed active participants.[58] Considering communication multimodal,[59] she documents ways in which non-human animals already participate in political and deliberative processes through their 'sheer presence', such as when the presence of newts or geese forms the grounds for resistance to an industrial development, or forces the relocation of a road. Equally there may be 'negotiation': for example, when companion animals negotiate their living arrangements with their human companions. She notes ways in which animal actions can in themselves constitute resistance, as when deer leave a territory to avoid the presence of hunters, or when captive animals organise to escape their cages or fields. These can be seen as direct forms of animal activism, in contrast with humans engaging in activism on animals' behalf. Meijer considers how interspecies communication is part of this: when captive animals communicate by gesture, sound or movement in escape, for instance, or when working animals communicate objection to oppressive conditions through gesture, sound or attacks on their handlers. She also highlights examples of political deliberation conducted through more-than-human communication.[60] For example, in the case of controversy around the use of land around Schiphol airport in Amsterdam, she interprets how the geese living there communicated their objections to attempts to move, capture or kill them through their flapping, sound and flight. Farmers responded by trying to make their land less attractive, to keep the geese away, and the airport authorities used dogs, solar panels and robot birds to scare them into changing their behaviour. The geese responded in various ways to these threats, after which strategies sometimes changed, and these interactions influenced the opinions of politicians involved with the site. Meijer is attentive to the ways existing political deliberations tilt the playing field so some can 'speak' (or be heard) more than others, according to race, age, background or ability. Yet recognising that, even amongst people, deliberative speech situations are never ideal opens the way to consider other dimensions to power—across human and non-human actors, for instance—and other kinds of deliberative communication that do not rely on human speech and language.

Such examples, where beings beyond the human communicate meaning and exercise political agency, can perhaps even extend to assemblages. When a river turns green with algae after pollution from surrounding industrial farmland, is this not a sign (as part of a set of possible multimodal signs, including the stench of death, and the absence of scents and sounds of life)—a metasign that expresses damage, even resistance? Those attuned to 'Mother Earth' might feel that it is, and farmers or environmental regulators may also pick up on the sign and respond (or not, turning a deaf ear) with changes to chemical inputs to which the assemblage of water, algae and chemicals will respond again, conveying further meanings through colour, density, flow. Such conversations are already happening; they are just not often seen to be conversations. We are accustomed to interpreting such instances as 'nature biting back' against human hubris or folly in tampering with or changing ecological systems inappropriately, through lack of understanding of their intricate dynamics, or perhaps in the pursuit of profit or power. Our focus in this book on communication in naturekind envisages a reworking of such examples, as instances, possibly, of non-human participation in deliberative environmental politics and activism.

Ike Sharpless has argued that insights from biosemiotics help understand how relations across the living world enter deliberative democracy and ecological activism, and probes how different animal ways of being, and the meanings they convey through signing processes embedded in their particular *umwelten*, allow for mutual recognition and a politics of joint action.[61] He suggests that what he terms 'transspecies pidgins' can bridge the worlds between beings with familiar and unfamiliar sensoriums, and that developing these would be significant in interspecies democracy. Amitav Ghosh, more accessibly perhaps, calls for a 'vitalist politics', in which the meanings shared between humans and non-humans are central. He cites the example of the long-standing protest movement against an oil pipeline in Dakota, which cuts across the territory of the Standing Rock Sioux Reservation, and is experienced there by many as a desecration of landscape. The campaign included storytelling, sweat lodges and rituals involving the river and other features of the landscape which became part of the campaign, playing roles as communicants and sharers of meaning.[62] Arts assist in such communication. Veronica Strang profiles how relations with water are condensed and rendered visible in anthropomorphic artistic expression across the world—as 'Mami Wata' figures in West Africa, for example—and are now appearing in protest movements and global environmental fora, and in doing so give 'the non-human world, and all of its inhabitants, a voice in the discussions and decisions that will shape the future of all living kinds'.[63] As packaged metasigns, they acquire signification in relation to new sign fields and evolving narrative genres.

These are not simply narratives by people about non-human natures and ecologies, but actually emerge out of communicative relations. As Amitav Ghosh writes bluntly, the idea that humans are the only storytelling animals, as so often assumed, has constituted another 'essential step in the silencing of nonhuman voices'.[64] Or, as Donna Haraway puts it, 'storying cannot any longer be put into the box of human exceptionalism'.[65]

As this book has expounded, the unfolding biological revolution is finally establishing that the phenomena that linguistics has been theorising for human language, that semioticians have been theorising for the human making of meaning and that anthropologists have been theorising around human culture all manifest beyond the human, across naturekind. These phenomena can be probed productively by extending the reasoning that has emerged from that part of naturekind to which we have privileged access—ourselves. Accordingly, as a political naturekind emerges, as it must, we shall find how more-than-human storying, and the metasigns and narrative structures in which meaning is made, are themselves emergent from what Michel Foucault cast as discursive relations that render stories inseparable from relations of power and their unfolding dialectics.

The environmental politics that will be necessary to secure a future in which human–non-human natures can thrive are also matters of emotion. Powerful protests and effective persuasion draw on the passion that emerges from communicative encounter. Care, conviviality and justice are matters of affect, not just matters to argue about. This book set out to integrate the new biology of communication, social learning and culture with long-standing theorisation in the social sciences and humanities that requires that attention be paid to meaning. With this as our focus, we open at once new paths to connectivities within and across naturekind that are deeply affective, emotive, emotional. Communication and affect are two sides of the same coin. Affect and the capacity to share meaning are built through encounters, however: a language must be developed. We cannot expect chickens and pigeons, forests, soils and seas, somehow to speak for themselves, or convey emotion in anything like the human ways to which we are accustomed; rather we must learn to pick up on and develop the signs and metasigns, syntax and metasyntax for communication. Those for whom developing such languages in companionable and immersive relations beyond the human is a familiar part of life, and who live on the margins of industrialised worlds where it is not, have much to teach. In writing this book, we hope to have drawn such experience into the limelight of environmental politics, as well as helping us all, perhaps, to recognise and enhance the conversations beyond the human in all our everyday lives. To bring such recognition to the surface and the centre is surely of crucial importance to the politics of now and the future: a politics of hope focused on connectivity, affect and communication across life in all its forms. This is a politics of, and for, naturekind.

NOTES

1. Introduction

1. The term 'pluriverse' encompasses the idea of many worlds within a world, of multiple ways of being (ontologies) as well as a plurality of visions and pathways of what might constitute positive change. See, for example, Blaser and de la Cadena (2018); Kothari et al. (2019).

2. Chapter 2 reviews these advances in the biological sciences.

3. Whiten (2017; 2019).

4. Bekoff and Pierce (2010); Briefer (2012); Filippi (2016); Filippi et al. (2017).

5. See, for example, Gagliano (2018); Nadkarni (2008); Sheldrake (2020).

6. No other term quite captures this; to refer to 'humans as part of nature', 'human relationships with nature' or 'nature-cultures' somehow implies the connection of separated kinds of life, reinforcing binaries. The notion of 'naturekind' should certainly not be understood as separate from 'humankind', or internally homogeneous.

7. Kohn (2013), esp. 8–9.

8. We discuss the development of biosemiotics and Sebeok's influence in detail in chapter 2. For a review, see Kull (2003). The distinction between linguistic and prelinguistic communication, finding commonality between humans and wider life in the latter, was also fundamental to the influential work of Gregory Bateson, who from the 1950s referred to symbolic dimensions as part of 'digital' communication associated with human language, in contrast with iconic and indexical dimensions that he considered as 'analog'. Bateson (1972; 1979); Guddemi (2020).

9. For example Crist (2019).

10. As Crist puts it, "The human supremacist worldview has never placed all humans on a par [. . . ;] so-called inferior humans were called beasts, savages and the like, and have endured humiliations, subjugations and genocides" (2019), 1.

11. Cornelli, McKirahan and Macris (2013).

12. Montaigne (1991 [1580]).

13. Curley (1994).

14. Morgan (2018 [1868]).

15. Haraway (2003; 2008).

16. Latour (1987; 2005).

17. Haraway (1991; 2016).

18. Deleuze and Guattari (1987).

19. Ingold (2011).

20. Bennett (2010).

21. Stengers (2010).

22. Barad (1996; 2007).

23. There are myriad such ethnographies, accumulating rapidly, and those cited here are just a small selection. Valuable collections include Ingold and Pálsson (2013); Kirksey (2014); and Fijn and Kavesh (2024). We refer to these and many others to explore focal issues in our later chapters, drawing out communicative dimensions left implicit or unaddressed in the original works.

24. Birke and Thompson (2018) (horses); Song (2010) and Van Dooren (2014) (birds); Tsing (2015) (mushrooms); de la Cadena (2015) (mountains).

25. Helmreich (2009; 2015) (oceans); Ogden (2011) (mangroves); Weston (2017) (urban landscapes).

26. See for example Castree and Nash (2004); Lorimer and Hodgetts (2024). Prominent works amongst many include Lorimer (2020); Swanson, Lien and Ween (2017); Barua (2023).

27. See, for example, Chibvongodze (2016); Etieyibo (2017); Kimmerer (2015); Watene (2024); and the discussion in Fraser et al. (2024).

28. See, for example, Nhemachena (2016); Nyamnjoh (2017; 2020); Todd (2016); Ghosh (2021); or indeed earlier ecofeminist works such as Merchant (1990).

29. Bateson (1972).

30. Despret and Buchanan (2016); see also Buchanan, Chrulew and Bussolini (2019).

31. Meijer (2019).

32. Lestel (2002); Moser (2022).

33. Lestel (1998), 203.

34. In the documentary *Nature and Us: A History Through Art* (2021), written and presented by James Fox for BBC 4, series 1, episode 1, minutes 11–12.

35. Multispecies ethnography, its expanding methodological repertoire and its challenges and limitations is itself a focus of recent works. See, for example, McLauchlan (2021); Ogden, Hall and Tanita (2013); Kirksey (2014).

36. See, for example, Purewal et al. (2019).

37. Tsing (2015); Weston (2017).

38. Soga and Gaston (2024). The term 'shifting baseline syndrome', since much elaborated, was originally coined by Pauly (1995).

39. Safina (2021).

40. Crist (2019).

41. Puig de la Bellacasa (2017), 2, who argues that care may be a particularly human trouble, but 'this does not make of care a human-only matter'.

2. Language and Culture Beyond the Human: Rethinking Communication Across Naturekind

1. This experiment is reported in the history of Robert Lindesay of Pitscottie, from the 1899–1931 edition by A.J.G. Mackay, Edinburgh: Wm. Blackwood for the Scottish Text Society, based on two of the oldest MSS, volume 1, p. 237. See Campbell and Grieve (1981).

2. Recent debates on language in the biological sciences are usefully reviewed by Berthet et al. (2023); Cartmill (2023); Prat (2019); Amphaeris et al. (2023).

3. Senthurran and Mason (2021); on vibrational communication across a wide range of species, see Cocroft et al. (2014).
4. Frisch (1967).
5. Hasenjager, Franks and Leadbeater (2022).
6. See, for example, Higham and Hebets (2013); Yong (2023).
7. Maynard-Smith and Harper (2003), 3.
8. Murray, Zeil and Magrath (2017).
9. Zhou et al. (2022).
10. Aubin, Jouventin and Hildebrand (2000).
11. Zuberbühler (2012).
12. Scott et al. (2023).
13. Worm, Landgraf and von der Emde (2021).
14. Jolly (1966).
15. Wohlleben (2016); Simard (2018). Whether the evidence is robust is considered by Karst, Jones and Hoeksema (2023).
16. Karban, Yang and Edwards (2013).
17. Halfwerk et al. (2019); see also Pouw et al. (2021).
18. Elias et al. (2012).
19. Pepperberg (2017).
20. Cissewski and Luncz (2021).
21. See the review in Knörnschild and Fernandez (2020).
22. Seyfarth, Cheney and Marler (1980).
23. Hersh et al. (2022).
24. Quick and Janik (2012).
25. Boesch (1991); Gabrić (2022).
26. Leroux et al. (2021); Leroux and Townsend (2020).
27. Arnold and Zuberbühler (2012).
28. Zuberbühler, Cheney and Seyfarth (1999); Suzuki, Wheatcroft and Griesser (2016).
29. Engesser, Ridley and Townsend (2016).
30. Suzuki, Wheatcroft and Griesser (2017), 2331.
31. Pleyer, Lepic and Hartmann (2022).
32. Bohn et al. (2008).
33. Searcy et al. (2022). See also Searcy, Chronister and Nowicki (2023).
34. Prat (2019).
35. See, for example, Bakker (2022); Mustill (2022).
36. Markowitz et al. (2013); see also Soma and Mori (2015).
37. Mann and Hoeschele (2020); Mann et al. (2021).
38. Allen et al. (2019).
39. Ferrigno et al. (2020).
40. Lameira et al. (2023).
41. Gentner et al. (2006).
42. Liao et al. (2022).
43. See, for example, Rey and Fagot (2023). This problem for hierarchy and recursion is discussed more generally by Cartmill (2023).

44. Schlenker et al. (2016); Ouattara, Lemasson and Zuberbühler (2009).
45. Bergman and Sheehan (2013); Wittig et al (2014).
46. Berthet et al. (2023).
47. Flower, Gribble and Ridley (2014).
48. Whitehead and Rendell (2015), 7.
49. Safina (2021), 50.
50. Whiten (2017); Galef and Whiten (2017); Whiten, Caldwell and Mesoudi (2016).
51. Wright (1996).
52. Smeele et al. (2024).
53. Marler and Tamura (1964).
54. Aplin (2019).
55. Prat et al. (2017).
56. Fernandez and Knörnschild (2020).
57. Laland, Atton and Webster (2011), 958.
58. Dong et al. (2023).
59. Danchin et al. (2018). Other important examples amongst the growing literature on insect social learning and culture include Reznkova (2023); Leadbeater and Chittka (2007); Grüter and Leadbeater (2015); Alem et al. (2016).
60. Witzany (2008), 60.
61. Karban, Grof-Tisza and Couchoux (2022).
62. Karban (2021); see also Karban (2015); Mancuso (2017); Trewavas (2014).
63. Whiten (2021).
64. See, for example, discussions in Amphaeris et al. (2023); Cartmill (2023); Hoeschele, Mann and Wagner (2023); Hoeschele, Wagner and Mann (2023).
65. Kull et al. (2009), 169; see also Wheeler (2016), who elaborates on the problems of this position, not least for the politics of life.
66. Wynne and Udell (2020), 14.
67. Morgan (1903), 59.
68. Wynne and Udell (2020), 4.
69. Feeley-Harnik (2021).
70. Taiz et al. (2019), 685. For accusations of mechanomorphism, see, for example, Crist (1999), arguing against Kennedy (1992). For a discussion finding common ground in making space for non-anthropomorphic representation, see, for example, Karlsson (2012).
71. 'Webs of meaning' is an allusion to the work of anthropologist Clifford Geertz (1973), accredited with the foundation of cultural anthropology. There are, of course, very many social science traditions grappling with the social shaping of culture and the place of meaning within it, ranging from ideas of collective consciousness in the work of Émile Durkheim to the historical-materialist traditions of Karl Marx, inflected by Antonio Gramsci, and the structuralist and post-structuralist traditions that we engage with below.
72. Albeit with an 'ecological revolution' over the last few decades drawing attention to complex, non-equilibrial dynamics and non-linearities, and to regularities as emergent, not given. See, for example, Levin (2005); Norberg (2004).
73. One long-standing ecological tradition does attend to human involvement, but often through mechanistic concepts such as 'social-ecological systems'. See, for example, Berkes and Folke (1998); Berkes, Colding and Folke (2003); Preiser et al. (2018).

74. Lestel (2014), 61; Moser (2022). Focusing just on human–animal relations, philosopher Mary Midgley (1983 once termed these 'mixed communities'.
75. Lestel cited in Moser (2022), 204.
76. See, for example, Despret (2015).
77. We are not alone in eschewing questions of mind and consciousness by focusing on communication. Gregory Bateson, for example, takes the focus off consciousness, arguing that much communication is below thresholds of 'conscious perception'. Equally, whilst he does develop a theory of mind, he does not locate this within individual organisms, but envisages mind as immanent in larger systems of which organisms are a part: see Bateson (1972), esp. 339.
78. Kull (1999), 386. The rapid emergence of biosemiotics as a discipline since the 1970s, its core assumptions and its multiple strands of debate, are captured in overviews and collections including Sebeok and Umiker-Sebeok (1992); Sebeok, Hoffmeyer and Emmeche (1999); Cobley et al. (2011); Cobley (2016); Emmeche and Kull (2011); Hoffmeyer (2008); Kull (2003); Barbieri (2008).
79. Augustine (1892), 35.
80. Augustine (1892), 35 (emphasis added).
81. Sebeok founded zoosemiotics which then developed into biosemiotics; see Kull (2003).
82. Although Sebeok developed this lineage of biosemiotics, his earliest work (e.g., Sebeok [1967]) does acknowledge the existence of arbitrary codes and conventions, which he studied in honeybees and then suggested applied to other animals. This may be related to his original intellectual grounding in cybernetics and the study of semiotic codes as being like digital codes.
83. 'What we know of zoosemiotic processes furnishes no evidence of syntactic structures, not even in any of the alloprimates.' Sebeok (1996), 108.
84. Thus Sebeok's *Perspectives in Zoosemiotics* found that 'subhuman species communicated by signs that appear to be most often coded analogically' (Sebeok [1972], 10) (and that he later termed 'indexical'), in contrast with human speech, which combined analogical and symbolic signs (the latter referred to as 'digital'). He followed up this review with a vast work examining the modalities of perception and communication in animals, in which he again claimed that there was, across taxa, a decisive role in animal behaviour of indexical signs (Sebeok [1977]). Animal life, it could thenceforth be said, was about 'the correct decipherment of indexical signs, ceaselessly barraging their *umwelt*'. Sebeok (1997), 282.
85. Sebeok argued that language emerged not through gradualist evolutionary development, but through evolutionary 'exaptation' (the process by which features acquire functions for which they were not originally adapted or selected: fortuitous by-products of human evolution). In positing radical mental discontinuity between humans and other species, such biosemioticians take the position of Alfred Russel Wallace *contra* Darwin.
86. Peirce became the 'lodestar' of biosemiotics and zoosemiotics, phytosemiotics, immunosemiotics and further sub-fields which all developed with their focus on the study of 'prelinguistic' signs. Kull (2003).
87. To understand the operation of 'prelinguistic' signs in the making of meaning, Peirce also outlined the need not just to consider the signifier and signified, but to invoke too the active role of a third interest, 'the interpretant', in semiosis. For prelinguistic signs there can be no externally derived convention, so the making of meaning requires there to be an interpreter. Thus in this move, we lose the need for a code, but gain the need for an interpreter. Peirce (1960 [1931–32]), 1:285.

88. Johansen (2002), 51.

89. Conceiving of some signing as 'indexical' or 'iconic' also attracted media analysts of the human world such as Gilles Deleuze, who sought to capture what they considered the more 'prelinguistic', direct and unmediated, 'embodied' communicative power of signs, for instance in film. They thought they could find such immediacy in indexical and iconic forms, as a supplement to 'symbolic' signs which they saw as too distanced and 'representational', the sign being always one step removed from experience.

90. Kull (2003).

91. Kohn (2013). This also converges with the emphasis on direct, sensuous communication of those taking a phenomenological approach to comprehend communication beyond the human, without labelling this as biosemiotics, such as philosopher David Abram and anthropologist Tim Ingold. Abram (2017 [1997]); Ingold (2000).

92. Such conceptual dominance has become institutionalised, such that non-Peircian, or post-Peircian, versions of biosemiotics are almost inconceivable. See Rodríguez Higuera (2020).

93. Toutain (2022).

94. Lévi-Strauss (1963 [1958]).

95. Saussure (2011 [1959]), 116.

96. Barthes (1967 [1964]), 9. Critiques were developed by Roman Jakobson in the 1960s and entered film and media studies in the 1980s, especially through the work of Gilles Deleuze. Jakobson (1965); Bogue (2003). Whilst Saussure envisaged linguistics as part of a general science of signs, Barthes construed all semiology as a part of Saussurean linguistics—which is broadly the position that we develop here.

97. Lévi-Strauss (1963 [1958]).

98. As Saussure put it, 'We should probably be unable to distinguish two ideas without the help of a language [...]. There are no: a) ideas already established and quite distinct from one another, b) signs for these ideas [...]. There is nothing at all distinct in thought before the linguistic sign.' Quoted in Toutain (2022), 103.

99. Deleuze (1986), ix.

100. See, for example, Welchman (2016).

101. Johansen (1988), 499.

102. Favareau (2010), 64; Toutain (2022) provides a critique. Whilst the idea of the 'pure' indexical sign and 'pure' iconic sign was alien to Peirce, it is currently fundamental to Peircean biosemiotics. Favareau even naturalises 'sign processes' as fundamental biological relations, conceiving of indexical, iconic or symbolic signs as being perceived within distinct neural circuits.

103. It was for this reason that Lévi-Strauss intuited that structural analysis would converge with theories of the subconscious.

104. Hutton (1999).

105. Lestel (2002), 55.

106. Kull (2010); Uexküll (2010 [1934]).

107. Greenough (2002), 337.

108. When geographer Henri Lefebvre captures how lived representational space is communicative and thus in some way enwrapped in life, he goes as far as to say that it is 'alive' and indeed might be said to speak and to command respect. Lefebvre (1991 [1974]), 42.

109. In the lexicon of biosemiotics, living beings establish a 'semiosphere': dialogues of meaning-making in which they create and reshape their own *umwelt* when interacting with the world around them in a 'functional circle'. To date, biosemiotics has restricted such analysis to (Peircean) 'natural' signs (Kull 1998).

110. For example, in a famous article on the Kabyle (Berber) house, Pierre Bourdieu showed how the shape, size, fabric and orientation of a building in relation to the sun is integral to the communicative order of those living in it—their 'habitus' and its structured sense of durability, normality. Bourdieu (1970).

111. Abram (2017 [1997]).

112. Ingold (2000), 409.

113. While beyond the scope of this book and unnecessary to its approach, debates on 'mind' are vast and long-standing. A popular account can be found in Ball (2022).

114. This also helps address what we might call the 'bee problem'; just as it is impossible to pin down an agentive 'mind' in either an individual bee or the collective hive, so it is impossible to pin down the origins of an idea to an individual person, or to the social and discursive world in which that individual is living. It is not a case of one or the other. Works grappling with similar challenges make use of notions like 'collective intelligence' or 'networked intelligence', the 'intelligence of a swarm', diffuse networks, and so on. See, for example, O'Bryan, Beier and Salas (2020).

115. The allusion here is to the Turing principle.

116. Douglas (2002 [1966]).

117. Bateson (1972), 275–76.

118. As Phillip Guddemi summarises it, 'It would never be possible to create, experimentally as it were, a communication without relational meta-implications of any sort.' Guddemi (2020), 4.

119. He did sometimes imply this, such as in his important work on interspecies play, considering, for example, how a dog's nip in the context of a playful interaction with an established carer is opposed to a bite in aggression towards an attacker; but his main interest here was in how negatives are conveyed in non-verbal communication, with no word 'not'. See Bateson (1972), 54–55.

120. Lakoff and Johnson (1980).

121. This is emphasised in the work of Roland Barthes, for whom packages build on other packages, accreting meanings and becoming layered through history into what he calls a metalanguage. We pick up this insight, but prefer the terminology of metasigns and metasyntax/narratives, retaining the distinction in semiotic analysis between the paradigmatic and syntagmatic axes. Whilst structural linguistic analysis focuses heuristically on the 'language system', as if a language could exist 'out of time'—'synchronically'—this downplays how language (and experience) are always historical in relation to the ways meanings become packaged, how packages sediment into established discourses, how packages relate to and might challenge each other and how all this is shaped by structural forces, and power.

122. In attending to the relations between structural power and meaning, Barthes's structural semiotic analysis finds common ground with what Michel Foucault analysed as 'discourse' and Karl Marx as 'ideology'. Foucault (1982); Eagleton (1991).

123. Cited in Campbell and Grieve (1981), 62–64.

124. Chomsky argued that language cannot be taken 'as a general model for the analysis of other realms of mind or culture (or as a specific manifestation of this)'. Cited in Keesing (1974), 251; see also Chomsky (2007).

125. Abram (2017 [1997]).

126. Attempts to categorise this variety taxonomically, such as by anthropologist Philippe Descola (Descola [2013]) are bold, but will always be flawed and reproduce the assumptions of those creating them. Neatly categorising framings risks singularising and stabilising something that is much more lived; rooted more in emergent experience than in philosophy.

127. This is the basis, for instance, of 'Amazonian perspectivism', as described by anthropologist Eduardo Viveiros de Castro, in which beings share similar interiority (are similar to us in this way) but take many different natures (physical forms). Viveiros de Castro (2014 [2009]).

128. Nelson (1983).

129. De la Cadena (2015).

130. Evans (1906); Carson (1917); Hyde (1916).

131. Leeson (2013).

132. Montaigne (1991 [1580]), 514.

133. Curley (1994).

134. Such influence can be detected in, for example, Deleuze and Guattari (1987) and Bennett (2010), albeit as part of otherwise different theoretical projects. Bennett, for instance, takes examples such as the grouping of thrown-down objects around a drain to highlight the way in which signing encompasses how things are arranged and how they move. In this 'an affective, speaking human body is not intrinsically different from the affective, signalling, nonhumans with which it coexists, hosts, enjoys, serves, consumes, produces and competes'. Bennett (2010), 117. These theoretical approaches have similarities, in turn, with the 'object-oriented ontologies' of those interested in the realities of non-living objects, independently of human perceptions of them: see Harman (2022).

135. Haraway (2003).

136. To put it another way, companionships can be viewed as involving intersubjective experiences that draw on only some, or a part, of each partner's multiple subjectivities. This echoes anthropologist Marilyn Strathern's argument that beings are not individuals, but 'dividuals', who establish 'partial connections' between parts of their own and other beings' divided selves. Strathern (1988).

137. See, for example, Wohllebehn (2016); Simard (2021); Figueiredo, Jens and Guggenberger (2021); all of whom develop network approaches in ecology in ways made possible by technological developments. See also Proulx, Promislow and Phillips (2005).

138. Attention to social networks has long been the stuff of social sciences and humanities, while 'network' is the preferred concept for some concerned with relations beyond the human too, such as the actor-networks of Bruno Latour and his followers that unite living and non-living beings, objects and even ideas. Latour (2005).

139. The concept of 'assemblage' evokes its own multistranded theories and debates; see, for example, Deleuze and Guattari (1987); Phillips (2006); Thiele (2016). Here we simply draw on key tenets of the approach.

3. Chickens

1. *Full Fact* (2020).
2. The evolution of chickens is much discussed, but studies point to origins in South-East Asia and chicken domestication being associated with the emergence of rice and millet cultivation. See Peters et al. (2022).
3. Roys Farm (2024), slightly adapted.
4. Marino (2017).
5. Bourdieu (1977 [1972]).
6. Lestel (2002).
7. Swancutt (2021).
8. Kockelman (2011), 444.
9. Tsegaye (2017).
10. Pachirat (2011).

4. Horses

1. Relations between humans and horses are the focus of a large literature across the biological and social sciences, and this chapter does not review or repeat this. Multispecies ethnographies are also proliferating, focusing on a range of activities and settings, from racing and archery to everyday equestrianism. Birke and Thompson (2018), Davis and Maurstad (2016) and Jones McVey (2023) provide reviews. Here we engage with these and others selectively from a communicative perspective and through a structural-biosemiotic lens.
2. Much of this is discussed in Birke and Thompson (2018), a collaboration between an ethologist and an anthropologist.
3. Jones McVey (2022), 470.
4. See Jones McVey (2022), 470; also Despret (2004; 2015).
5. Pony Club (2009 [1950]), 30.
6. Birke and Thompson (2018), 15.
7. Anthropologist Rosalie (or Rosie) Jones McVey reveals many of these in her popular review of horse riding and training traditions across the world, both formalised and less formalised. Jones McVey (2015).
8. Birke (2007).
9. Jones McVey (2022); on the role of riding lessons in this, see also Dashper (2016).
10. For a discussion of performance in human–horse relations, engaging with Judith Butler's notion of human gender relations as performed, see Birke, Bryld and Nikke (2004).
11. Maurstad, Davis and Cowles (2013), 328.
12. Despret (2004), 115.
13. Ingold (2000).
14. Jones McVey (2015).
15. See, for example, Pearce (2019).
16. Haraway (2003); Maurstad, Davis and Cowles (2013).
17. Jones McVey (2022).
18. Maurstad, Davis and Cowles (2013).

19. Game (2001), 3; see also Ford (2019).
20. Jones McVey (2022).
21. Argent (2012).
22. Birke and Thompson (2018), 15.
23. Fijn (2021), 69.
24. Irvine (2014), 88.
25. Peemot (2017).

5. Plants

1. The reasoning behind this argument concerning communication without mind in the vegetal world is not far from that of anthropologist Eduardo Kohn in his *How Forests Think* and philosopher Michael Marder's *Plant Thinking*, construing how plants 'think without a head', in a non-cognitive mode, enabling forms of 'non-conscious intentionality' that unfold actively through the multidirectional relationships between plants, other beings and their surroundings. Both scholars, however, theorise this in non-linguistic terms, referring, whether implicitly or explicitly, either to the so-called direct signing processes that Peirce identifies as iconic or indexical, or to more 'vitalist' conceptions of nature. Kohn (2013); Marder (2013a), 125. See also Marder (2013b).
2. Karban (2021); Witzany (2008).
3. Simard (2021); Wohlleben (2016).
4. Wohlleben (2016), 7.
5. Wohlleben (2016), 7.
6. Simard (2021).
7. Wohlleben (2016), 7.
8. Marder (2013a), 125.
9. Gibson (2018), 92.
10. Multispecies ethnographers are considering the relational agency of plants in numerous settings. Examples include relations with gardeners in urban Toronto and northern England, medicinal plant gatherers in southern Africa and spiritual herbalists in southern California. Elton (2021); Degnen (2009); Gibson and Ellis (2018); Vine (2019).
11. Phillips and Schulz (2021).
12. Phillips and Schulz (2021), 380.
13. Phillips and Schulz (2021), 382.
14. Lasco (2020).
15. Archambault (2016), 30.
16. Vine (2019).
17. Miller (2019).
18. Such contexts include research; as multispecies ethnographer Gibson observed when studying medicinal plants in southern Africa, 'We paused, spent time to dwell with plants, drew on all our senses and learned new ways to be attentive to plants and their more-than-human sociality and entanglements.' Gibson (2018), 92.
19. Degnen (2009), 162–63.
20. McKenzie (2022).

6. Bees

1. Frisch (1967).
2. Seeley (2010), 4.
3. Hölldobler and Wilson (2009), 6, 84.
4. Seeley (2010).
5. Crist (2004).
6. Crist (2004), 14.
7. Gould and Gould (1995 [1988]), 36.
8. Gould and Gould (1995 [1988]), 99.
9. Seeley (2010).
10. Singla (2020).
11. Dong et al. (2023).
12. Leadbeater and Chittka (2007); Oliveira and Bshary (2021).
13. Ball (2022).
14. See Crist (2004) for an interesting discussion of this dilemma and how it has been negotiated within the scientific community studying bees.
15. Lindauer (1971 [1961]), 59.
16. Grüter and Czaczkes (2019).
17. Piqueret, Sandoz and d'Ettorre (2019).
18. Sebeok (1995), 121; see also Sebeok (1967).
19. Peter Cowin: see https://www.beewhisperer.us (accessed 26 January 2025).
20. *PerfectBee* (n.d.).
21. Crist (2004), 24–25.
22. Piva (2023).
23. Some other bee and human senses are less aligned—such as touch (when honeybees use their antennae to identify each other, an ability denied to antennae-less humans) and pheromone sense (when each queen produces her own pheromone, inhibiting other females from laying eggs and drawing her brood to her, which is beyond human capacity to pick up on with any sensitivity).
24. Cortopassi-Laurino et al. (2006), 276.
25. Piva (2023), 113.
26. Sebeok similarly noted the sophistication of the Meliponini guide bee and odour trail, as well as the fact that, whereas the waggle dance communicates meanings along a horizontal axis, the odour trail can also go up and down vegetation, conveying meanings along a vertical axis. Sebeok (1967), 71.
27. Piva (2023), 115.
28. Moore and Kosut (2013).
29. Moore and Kosut (2013), 520.

30. Moore and Kosut (2013), 528.
31. Novellino (2019).
32. Novellino (2019), 193.
33. Baptista (2022), 41.
34. Casper and Moore (2009).
35. Monchanin et al. (2021).

7. Bats

1. Sieradski and Mikkola (2022), who note that early classifications divided bats based on morphology and behaviour into two suborders, *Microchiroptera* (microbats) and *Megachiroptera* (flying foxes and Old World fruit bats), although more recent molecular biology indicates two further suborders, *Yinpterochiroptera* and *Yangochiroptera*.
2. Wynne and Udell (2020), 34, 35.
3. Nagel (1974).
4. Kull (2010).
5. See, for example, discussion in Mac Aodha et al. (2018).
6. Borelli (2013).
7. Ayivor et al. (2017); Ohemeng et al. (2017).
8. Ohemeng et al. (2017), 186.
9. As described, for instance, by Ayivor et al. (2017).
10. Ohemeng et al. (2017). See also Cros (2020).
11. Olival (2016), 1.
12. Kamins, Rowcliffe and Restif (2014); see also Boyales et al. (2011).
13. Neer (2013).
14. Sieradzki and Mikkola (2022).
15. Jeffreys (1944).
16. See White (2000).
17. Bambi (2017).
18. Amongst many possible sources, see, for example, Schneeberger and Voigt (2016).
19. Leroy et al. (2005).
20. Chua et al. (2000).
21. Luby et al. (2006).
22. Leach et al. (2022).
23. See, for example, Chen, Hwang and Shaw (2023). Such accounts draw on research and modelling studies that purport to link the spatial distribution of Ebola outbreaks with trends in deforestation and forest fragmentation, such as Olivero et al. (2017) and Rulli et al. (2017).
24. Walker and Nadin (2011).
25. *Born Free* website (http://www.bornfree.org.uk). 'Proposal to Cull Uganda's Wildlife', 7 November 2007 (accessed 27 June 2016; page no longer available).
26. A village narrative that seems entirely plausible in the light of evidence from Guinea's 2021 outbreak that the Ebola virus can 'hide' in the human body and then resurge and be transmitted by human 'survivors'. See Fairhead and Millimouno (2017); Fairhead, Leach and Millimouno (2021).

27. See Luby et al. (2006); Nahar et al. (2014).

28. One Health approaches themselves are varied and emergent; for a recent review and contemporary framing, see, for, example Zinsstag et al. (2023). Our arguments here align more closely with locally grounded approaches, as described in Leach et al. (2017).

29. Johnson et al. (2020), 1.

30. Souris et al. (2022); see also Borrega et al. (2021). There are also other plausible explanations for the so-called African Paradox, including demographic ones (very young populations linked to long-standing health inequalities and low life expectancy); socio-economic ones (the predominance of open-air livelihoods and living), and possibly political ones (strong early leadership in control measures). The likelihood is that several explanations interact.

31. Lorimer (2020).

32. Sagan (2013).

33. Johnson et al. (2018).

8. Forests

1. FAO (2020).

2. Kohn (2013).

3. Kohn (2013), 42. A not dissimilar broad conceptualisation of forest animal life in Gabon as a mass of 'thinking' interactions was aired at the dawn of the twentieth century: see Du Chaillu (1900).

4. Forests are a major focus in the recent biological science of communicative networks, as reviewed in works aimed also at a popular audience, such as Sheldrake (2020), Simard (2021), Wohlleben (2016) and Gagliano (2018).

5. Wohlleben (2016).

6. Kimmerer (2015).

7. Kimmerer (2015), 16.

8. Kimmerer (2015), 19–20.

9. Kimmerer (2015), 20.

10. Kimmerer (2015), 56.

11. We have explored these processes and relations ethnographically in the forest-savanna mosaic landscape of Kissidougou Prefecture, Guinea, in Fairhead and Leach (1996), related articles and in the film *Second Nature*, and have extended the analysis and its implications to the wider West African region and beyond in Fairhead and Leach (1995; 1996; 1998; 2003b). Whilst these works were not originally conceived through a communicative multispecies lens, the experiences they document can aptly be revisited in this way.

12. Kimmerer (2015).

13. See, for example, discussions and exemplification of what anthropologists have come to call 'Amazonian perspectivism' in Viveiros de Castro (2014).

14. Fairhead and Leach (1996; 1998).

15. Rosevear (1979), 78 (emphasis added).

16. For example, https://forestbathingsussex.co.uk/ (accessed 26 January 2025). Those promoting forest immersions promote practices that draw on diverse therapeutic traditions—this

particular example claims alignment with Swedish yoga and the Japanese awareness meditation *shinrin-yoku*—but all share the notion that being in forest has beneficial effects.

17. Kimmerer (2015), 21–22.
18. See, for example, Leach and Scoones (2015b).

9. Soils

1. For a review, see Barrera-Bassols and Zinck (2003).
2. Krzywoszynska and Marchesi (2020), 194, and articles in the special journal section of which this is a part; see also Salazar et al. (2020).
3. See, for example, the review in Bender, Wagg and van der Heijden (2016).
4. Puig de la Bellacasa (2015), 701.
5. Puig de la Bellacasa (2015), 701. For discussions of the complex history and contested present of such perspectives within soil science and ecology, and their intertwining with processes of social and economic change, see Hartemink (2016); Warkentin (2006); Marchesi (2020).
6. Sheldrake (2020).
7. Monbiot (2023), 16.
8. Multispecies ethnographies of living soils reconfigure humans as full participant members of a soil community, rather than simply producers or consumers of its products, yet underplay their communicative dimensions. See, for example, Krzywoszynska (2020) in a UK context, and O'Brien (2020) on pasture-cropping in Australia.
9. Puig de la Bellacasa (2017), 147.
10. For a more detailed discussion, see Fairhead and Leach (1996; 2003a).
11. Owens and Rutledge (2005).
12. We have elaborated in depth on the social, ecological and soil science dimensions of these 'anthropogenic dark earths' elsewhere; here we very briefly revisit this work from a communicative angle. For further detail, see Fraser, Leach and Fairhead (2014); Frausin et al. (2014); Solomon et al. (2016).
13. An ethnopedological study in the humid tropical region of Côte d'Ivoire similarly noted that people distinguish between dark, light and red soils: Birmingham (2003).
14. Frausin et al. (2014), 13.
15. Glaser and Birk (2012).
16. For instance, twelve major soil texture types are defined by the United States Department of Agriculture according to relative mineral proportions: sand, loamy sand, sandy loam, loam, silt loam, silt, sandy clay loam, clay loam, silty clay loam, sandy clay, silty clay and clay. USDA (n.d.).
17. Fairhead and Leach (1996); for a further discussion in the context of Sierra Leone, see Boone (1990).
18. See, for example, Scoones (2015).
19. Puig de la Bellacasa (2015; 2017).
20. HLPE (2019).
21. Quoted by Green Connect (2020).
22. Green Connect (2020).
23. Quoted in Puig de la Bellacasa (2015), 701.

24. Krzywoszynska, Banwart and Blacker (2020); Krzywoszynska (2019).
25. O'Brien (2020), 270.
26. As documented, for instance, in Shiva (2008).
27. Biehl and Staudenmeier (1995).
28. Biehl and Staudenmeier (1995), 13.
29. Moore-Colyer (2004).
30. Ghosh (2021), 223; see also Winn (2002).
31. Marya and Patel (2021).
32. This is the argument put forward, for instance, by Monbiot (2023), although it is much contested—not least by farmers for whom it signals an end to their future.
33. Ambitious schemes to sequester (and commoditise) soil carbon have become part of the set of policy responses to global climate change; see, for example, Lefèvre et al. (2017) and Minasny et al. (2017).

10. Seas

1. Our focus on communication adds distinctively to a so-called 'oceanic turn' in the environmental humanities and social sciences, as well as to a 'maritime anthropology' that is as old as the discipline itself. Recent works across many settings explore human–non-human relations in and with seas, often including their material and technological elements: how the qualities, mobility and mutability of seawater are figured across the world, as well as how seas provide metaphors and lenses through which to reflect on human social life. We draw selectively here on this large literature. For overviews, see Moore and Jacka (2020); Deloughrey (2017); Johnson (1996); Helmreich (2011).
2. Strang (2023), 1.
3. As Helmreich (2023) explores, waves themselves are open to multiple meanings and framings across various sciences and societies.
4. Meanings and practices of surfing are explored in a growing range of ethnographies from around the world. See overviews in Ford and Brown (2006) and the discussion of 'surf cultures' in Helmreich (2023).
5. Whyte (2019), 13.
6. Whyte (2019), 20.
7. Anderson (2013), 957.
8. Kerby (2010), 43.
9. Anderson (2013), 958.
10. Kerby (2010).
11. Anderson (2013).
12. SAS (2025).
13. See, for example, https://www.bbc.co.uk/news/uk-england-46836867 (accessed 9 February 2025).
14. Anderson (2012).
15. Similarly, technologies used by scientists and researchers become part of the assemblages through which they relate to the sea, and imbricated in their particular framings that follow what we would see as narrative genres. See, for example, Helmreich (2009).

16. Safina (2021).
17. For a compelling overview of this part of the biological revolution, see Mustill (2022).
18. Morell (2023).
19. Hayward (2010).
20. Zerner (2003).
21. Sea Change Trust (2024).
22. *My Octopus Teacher*, a 2020 Netflix original documentary film directed by Pippa Ehrlich and James Reed.
23. Frankal (2020), 3.
24. Keane (2011).
25. Yap and Yu (2019).
26. Walley (2004).
27. Strang (2023), 31.
28. Sabo and Sabo (1985), cited in Strang (2023), 82.
29. Leach (1994).
30. For example, Lauren Drakopulos traces how human, animal and non-material entities are assembled in industrial fishing in ways that construct and support the concept of 'bycatch', depicting dolphins and sea turtles captured in fishing nets as an unintended consequence of fishing. Drakopulos (2020).
31. Pálsson (1998).
32. Helmreich (2011).
33. Christine Walley, for instance, shows how people of Tanzania's Mafia Island did not share the distinction between nature and culture to which transnational marine scientist park consultants were committed: far from being a wild 'other' to human culture, their seascape was a space of fishing, fish and seafaring stories. Walley (2004).
34. Pauly (1995).

11. Cities

1. Data from *Demographia* (2023), 2 and UN DESA (2018) respectively
2. Debates over the framing of cities as non-living vs. living form part of a vast literature in urban studies that we cannot review here. Recent social science works taking a 'more-than-human' approach to living cities consider them variously as 'cosmopolitical' assemblages of human and non-human actors and networks (Farías and Blok [2016]) or as emergent outcomes of interactions between a multitude of entities and beings (Barua [2023]).
3. See Kennedy, Cuddihy and Engel-Yan (2007).
4. Lisdorf (2021).
5. Price (2018).
6. Barua (2023); Hinchliffe and Whatmore (2006).
7. Bennett (2010), 5.
8. See, for example, Rydin (2013).
9. Hinchliffe and Whatmore (2006).
10. Baldaccini et al. (2000).
11. Skandrani, Desquilbet and Prévot (2018).

12. Wynne and Udell (2020), 172–74.
13. Wynne and Udell (2020), 177.
14. Watanabe et al (1995).
15. Skandrani et al. (2018), 281.
16. Kavesh (2021), 45.
17. Kavesh (2021), 48.
18. Barua et al. (2021).
19. Barua et al. (2021), 2.
20. See, for example, Allen et al. (2014).
21. For example, World Health Organization (2016); Wendelboe-Nelson et al. (2019).
22. Fitzgerald et al. (2019).
23. Peen et al. (2010); Paykel et al. (2000); Coward et al. (1983).
24. Wendelboe-Nelson et al. (2019).
25. Nutsford, Pearson and Kingham (2013).
26. Van den Berg et al. (2010).
27. Barton and Pretty (2010); Shanahan et al. (2015).
28. White et al. (2019).
29. Barua et al. (2021), 3.
30. Bister, Klausner and Niewohner (2016).
31. Barua et al. (2021), 6.
32. See https://www.ebonyhorseclub.org.uk/ (accessed 27 January 2025).
33. Dobson et al. (2021).
34. Dobson et al. (2021), 5.

12. Conclusion: A Political Naturekind

1. Ghosh (2021), 197.
2. Haraway (2016).
3. Haraway (2003), 14.
4. Lestel and Taylor (2013), 183 (original emphasis).
5. Mbembe (2019).
6. Slade and Alleyne (2023).
7. Arendt (1963), 272.
8. Nadasdy (2007, 25).
9. See, for example, accounts in Leach (2000).
10. Iamblichus (1818), 134.
11. Iamblichus (1818), 79.
12. Tsing (2015); Weston (2017).
13. Fairhead, Leach and Scoones (2012).
14. Marris (2013).
15. Buscher and Fletcher (2020).
16. Todd (2016).
17. See, for example, Colchester (2004); Berkes (2007).
18. Fitz-Henry (2022).

19. Hinchliffe and Whatmore (2006), 124.
20. Jepson (2019), 123.
21. See DEFRA (2024); Richardson, Dobson and Abson (2020).
22. McAfee (1999).
23. Tronto (2020), 103.
24. Puig de la Bellacasa (2017).
25. Scoones et al. (2020).
26. Crist (2019), 163.
27. Classic texts in the multi-stranded movement that is ecofeminism include Mies and Shiva (1993), Warren (2000) and Plumwood (2003), whilst feminist political ecology addresses material and knowledge dimensions of intersecting feminist and ecological struggles: see, for example, Harcourt and Nelson (2015).
28. Celermajer et al. (2021); Chao, Bolender and Kirksey (2022); Lamara et al. (2024).
29. In this, the concerns of environmental justice (which has conventionally addressed relations of social justice among people) and 'ecological' justice (conventionally concerned with the non-human world) are merging. See Schlosberg (2004), and Dryzek and Pickering (2018), 67, who propose the term 'planetary justice' to encompass multispecies justice at scales up to the global.
30. Dryzek and Pickering (2018), 71.
31. Celermajer et al. (2021), 7.
32. These tensions are a matter of debate that flared up, for instance, over the 'Half Earth' movement, with Bram Buscher and colleagues arguing that this risks treating humans unjustly, by protecting for nature land that people depend on for their livelihoods, whilst Helen Kopnina argues that it is a necessary way to treat non-humans justly. Buscher et al. (2017); Kopnina (2016).
33. Fitz-Henry (2022).
34. Pearce (2010); Raworth (2017).
35. Fairhead (2018).
36. Patterson et al. (2017).
37. Stirling (2015).
38. See also Scoones, Leach and Stirling (2015); Scoones et al. (2020).
39. Petrović (1991).
40. See Lenin (2008); Gramsci (1971).
41. Polanyi (2001 [1944]).
42. Arendt (1998 [1958]).
43. Fraser (2013).
44. Leach, Scoones and Wynne (2005); Leach and Scoones (2015a).
45. Tilly and Tarrow (2015).
46. Comyn (1922), 165.
47. Habermas (1984 [1981]). Indeed, this has been true of our own earlier contributions to this field: for example, Leach, Scoones and Stirling (2010); Fairhead and Leach (2003b).
48. Meijer (2019), 151.
49. Meijer (2019).
50. Youatt (2020).

51. Latour (2005; 2018).
52. Stengers (2010)
53. Latour (1999), 290.
54. Hinchliffe and Whatmore (2006), 17.
55. Dryzek and Pickering (2018), 17.
56. Dryzek and Pickering (2018), 126.
57. Macy and Brown (2014).
58. Meijer (2019), who also reviews related work on animal activism.
59. Many of Meijer's insights are akin to those in biosemiotics, although oddly, her otherwise wide-ranging view of animal languages does not refer specifically to the discipline or engage with its literatures.
60. Meijer (2019), 177, where she presents a case study of 'Goose Politics' involving resistance, deliberation and the politics of space.
61. Sharpless (2016).
62. Ghosh (2021), 239.
63. Strang (2023), 15.
64. Ghosh (2021), 201.
65. Haraway (2016), 39.

REFERENCES

Abram, D. 2017 [1997]. *The Spell of the Sensuous: Perception and Language in a More-than-Human World*. New York: Vintage Books.
Alem, S., C. J. Perry, X. Zhu et al. 2016. 'Associative Mechanisms Allow for Social Learning and Cultural Transmission of String Pulling in an Insect'. *PLOS Biology* 14: e1002564.
Allen, J., R. Balfour, R. Bell and M. Marmot. 2014. 'Social Determinants of Mental Health'. *International Review of Psychiatry* 26(4): 392–407.
Allen, J. A., E. C.Garland, R. A. Dunlop and M. J.Noad. 2019. ' Network Analysis Reveals Underlying Syntactic Features in a Vocally Learnt Mammalian Display, Humpback Whale Song'. *Proceedings of the Royal Society B: Biological Sciences* 286: 20192014.
Amphaeris, J., D. T. Blumstein, G. Shannon, T. Tenbrink and A. Kershenbaum. 2023. 'A Multifaceted Framework to Establish the Presence of Meaning in Non-Human Communication'. *Biological Reviews* 98: 1887–1909.
Anderson, J. 2012. 'Relational Places: The Surfed Wave as Assemblage and Convergence'. *Environment and Planning D: Society and Space* 30(4): 570–87.
Anderson, J. 2013. 'Cathedrals of the Surf Zone: Regulating Access to a Space of Spirituality'. *Social & Cultural Geography* 14(8): 954–72.
Aplin, L. M. 2019. 'Culture and Cultural Evolution in Birds: A Review of the Evidence'. *Animal Behaviour* 17: 179–87.
Archambault, J. 2016. 'Taking Love Seriously in Human-Plant Relations in Mozambique'. *Cultural Anthropology* 31(2): 244–71.
Arendt, H. 1963. *Eichmann in Jerusalem: A Report on the Banality of Evil*. New York: Viking Press.
Arendt, H., 1998 [1958]. *The Human Condition*. Chicago: University of Chicago Press.
Argent, G. 2012. 'Toward a Privileging of the Nonverbal: Communication, Corporeal Synchrony, and Transcendence in Humans and Horses'. In *Experiencing Animal Minds: An Anthology of Human-Animal Encounters*, edited by J. A. Smith and R. W. Mitchell, 111–28. New York: Columbia University Press.
Arnold, K. and K. Zuberbühler. 2012. 'Call Combinations in Monkeys: Compositional or Idiomatic Expressions?'. *Brain and Language* 120(3): 303–9.
Aubin, T., P. Jouventin and C. Hildebrand. 2000. 'Penguins Use the Two-Voice System to Recognize Each Other'. *Proceedings of the Royal Society B: Biological Sciences* 267(1448): 1081–87.
Augustine (of Hippo). 1892. 'On Christian Doctrine', translated by J. F. Shaw. In *The Works of Aurelius Augustine, Bishop of Hippo: A New Translation*, edited by Marcus Dods, vol. 9: 'On Christian Doctrine'; 'The Enchiridion'; 'On Catechising'; 'On Faith and the Creed', 1–171. Edinburgh: T. & T. Clark.

Ayivor, J. S., F. Ohemeng, E. Tweneboah Lawson, L. Waldman, M. Leach and Y. Ntiamoa-Baidu. 2017. 'Living with Bats: The Case of Ve Golokuati Township in the Volta Region of Ghana'. *Journal of Environmental and Public Health* 2017: 5938934.

Bakker, K. 2022. *The Sounds of Life: How Digital Technology Is Bringing Us Closer to the Worlds of Animals and Plants*. Princeton, NJ: Princeton University Press.

Baldaccini, N. E., D. Giunchi, E. Mongini and L. Ragionieri. 2000. 'Foraging Flights of Wild Rock Doves (*Columba l. livia*): A Spatio-Temporal Analysis'. *Italian Journal of Zoology* 67(4): 371–77.

Ball, P. 2022. *The Book of Minds: Understanding Ourselves and Other Beings, From Animals to Aliens*. Chicago: University of Chicago Press.

Bambi, J. 2017. 'Malawi "Vampire" Rumours and Hunt'. *Africanews*: 'The Morning Call', 30 October. https://www.africanews.com/2017/10/30/malawi-vampire-rumours-and-hunt-the-morning-call// (accessed 26 January 2025).

Baptista, J. A. 2022. 'Bodyland'. *American Ethnologist* 49: 35–39.

Barad, K. 1996. 'Meeting the Universe Halfway: Realism and Social Constructivism without Contradiction'. In *Feminism, Science, and the Philosophy of Science*, edited by L. H. Nelson and J. Nelson, 161–94. Dordrecht: Kluwer Academic.

Barad, K. 2007. *Meeting the Universe Halfway: Quantum Physics and the Entanglement of Matter and Meaning*. Durham, NC: Duke University Press.

Barbieri, M., ed. 2008. *The Codes of Life: The Rules of Macroevolution*. Berlin: Springer.

Barrera-Bassols, N. and J. A. Zinck. 2003. 'Ethnopedology: A Worldwide View on the Soil Knowledge of Local People'. *Geoderma* 111(304): 171–95.

Barthes, R. 1967 [1964]. *Elements of Semiology*, translated by A. Lavers and C. Smith. London: Jonathan Cape.

Barton, J. and J. Pretty. 2010. 'What Is the Best Dose of Nature and Green Exercise for Improving Mental Health? A Multi-Study Analysis'. *Environmental Science and Technology* 44: 3947–55.

Barua, M. 2023. *Lively Cities: Reconfiguring Urban Ecology*. Minneapolis: Minnesota University Press.

Barua, M., S. Jadhav, G. Kumar, U. Gupta, P. Justa, and A. Sinha. 2021. 'Mental Health Ecologies and Urban Wellbeing'. *Health and Place* 69(8): 102577.

Bateson, G. 1972. *Steps to an Ecology of Mind*. Chicago: University of Chicago Press.

Bateson, G. 1979. *Mind and Nature: A Necessary Unity*. New York: Bantam Books.

Bekoff, M. and J. Pierce. 2010. *Wild Justice: The Moral Lives of Animals*. Chicago: University of Chicago Press.

Bender, F., C. Wagg and M. van der Heijden. 2016. 'An Underground Revolution: Biodiversity and Soil Ecological Engineering for Agricultural Sustainability'. *Trends in Ecology and Evolution* 31(6): 440–52.

Bennett, J. 2010. *Vibrant Matter: A Political Ecology of Things*. Durham, NC: Duke University Press.

Bergman, T. J. and M. J. Sheehan. 2013. 'Social Knowledge and Signals in Primates'. *American Journal of Primatology* 75: 683–94.

Berkes, F. 2007. 'Community-Based Conservation in a Globalized World'. *PNAS* 104(39): 15188–93.

Berkes, F., J. Colding and C. Folke. 2003. *Navigating Social-Ecological Systems: Building Resilience for Complexity and Change*. Cambridge: Cambridge University Press.

Berkes, F. and C. Folke. 1998. *Linking Social and Ecological Systems: Management Practices and Social Mechanisms for Building Resilience*. Cambridge: Cambridge University Press.

Berthet, M., C. Coye, G. Dezecache and J. Kuhn. 2023. 'Animal Linguistics: A Primer'. *Biological Reviews* 98: 81–98.

Biehl, J. and P. Staudenmeier. 1995. *Ecofascism Revisited: Lessons from the German Experience*. Porsgrunn: New Compass Press.

Birke, L. 2007. 'Learning to Speak Horse: The Culture of "Natural Horsemanship"'. *Society and Animals* 15(3): 217–39.

Birke, L., M. Bryld and N. Nikke. 2004. 'Animal Performance: An Exploration of Intersections Between Feminist Science Studies and Studies of Human–Animal Relationships'. *Feminist Theory* 5: 167–83.

Birke, L. and K. Thompson. 2018. *Unstable Relations: Horses, Humans and Social Agency*. London: Routledge.

Birmingham, D. M. 2003. 'Local Knowledge of Soils: The Case of Contrast in Côte d'Ivoire'. *Geoderma* 111(3–4): 481–502.

Bister, M. D., M. Klausner and J. Niewohner. 2016. 'The Cosmopolitics of "Niching": Rendering the City Habitable Along Infrastructures of Mental Health Care'. In *Urban Cosmopolitics: Agencements, Assemblies, Atmospheres*, edited by I. Farías and A. Blok, 187–206. London: Routledge.

Blaser, M., and M. de la Cadena. 2018. 'Introduction: Pluriverse; Proposals for a World of Many Worlds'. In *A World of Many Worlds*, edited by. M. de la Cadena and M. Blaser, 1–22. Durham, NC: Duke University Press.

Boesch, C. 1991. 'Symbolic Communication in Wild Chimpanzees?' *Human Evolution* 6(1): 81–89.

Bogue, R. 2003. *Deleuze on Cinema*. London: Routledge.

Bohn, K., B. Schmidt-French, S. Ma and G. Pollak 2008. 'Syllable Acoustics, Temporal Patterns, and Call Composition Vary with Behavioral Context in Mexican Free-Tailed Bats'. *The Journal of the Acoustical Society of America* 124(3): 1838–48.

Boone, S. A. 1990. *Radiance from the Waters: Ideals of Feminine Beauty in Mende Art*. New Haven, CT: Yale University Press.

Borelli, L. M. 2013. 'Ghana: The Bats Who Never Left Their Chief'. *Farm Radio Weekly*, 26 May.

Borrega, R., D.K.S Nelson, A. P.Koval et al. 2021. 'Cross-Reactive Antibodies to SARS-CoV-2 and MERS-CoV in Pre-COVID-19 Blood Samples from Sierra Leoneans'. *Viruses* 13(11): 2325.

Bourdieu, P. 1970. 'The Berber House or the World Reversed'. *Social Science Information* 9(2): 151–70.

Bourdieu, P. 1977 [1972]. *Outline of a Theory of Practice*, translated by Richard Nice. Cambridge: Cambridge University Press.

Boyales, J. G., P. M. Cryan, G. F. McCracken and T. H. Kunz. 2011. 'Economic Importance of Bats in Agriculture'. *Science* 332: 41–42.

Briefer, E. F. 2012. 'Vocal Expression of Emotions in Mammals: Mechanisms of Production and Evidence'. *Journal of Zoology* 288(1): 1–20.

Buchanan, B., M. Chrulew and J. Bussolini, eds. 2019. *The Philosophical Ethology of Vinciane Despret*. London: Routledge.

Buscher, B. and R. Fletcher. 2020. *The Conservation Revolution: Radical Ideas for Saving Nature Beyond the Anthropocene*. London: Verso.

Buscher, B., R. Fletcher, D. Brockington et al. 2017. 'Half-Earth or Whole Earth? Radical Ideas for Conservation, and Their Implications'. *Oryx* 51(3): 407–10.

Campbell, R. and R. Grieve. 1981. 'Royal Investigations of the Origin of Language'. *Historiographia Linguistica* 9(1/2): 43–74.

Carson, H. L. 1917. 'The Trial of Animals and Insects: A Little Known Chapter of Mediæval Jurisprudence'. *Proceedings of the American Philosophical Society* 56(5): 410–15.

Cartmill, E. A., 2023. 'Overcoming Bias in the Comparison of Human Language and Animal Communication'. *PNAS* 120(47): e2218799120.

Casper, M. J. and L. J. Moore. 2009. *Missing Bodies: The Politics of Visibility*. New York: New York University Press.

Castree, N. and C. Nash, eds. 2004. 'Mapping Posthumanism: An Exchange'. *Environment and Planning A: Economy and Space* 36(8): 1341–63.

Celermajer, D., D. Schlosberg, L. Rickards et al. 2021. 'Multispecies Justice: Theories, Challenges, and a Research Agenda for Environmental Politics'. *Environmental Politics* 30(1–2): 119–40.

Chao, S., K. Bolender and E. Kirksey, eds. 2022. *The Promise of Multispecies Justice*. Durham, NC: Duke University Press.

Chen, C., I. Hwang and A. Shaw. 2023. 'On the Edge: The Next Deadly Pandemic is Just a Forest Clearing Away. But We're Not Even Trying to Prevent It'. *ProPublica*, 27 February. https://www.propublica.org/article/pandemic-spillover-outbreak-guinea-forest-clearing (accessed 26 January 2025).

Chibvongodze, D. T. 2016. 'Ubuntu is Not Only About the Human! An Analysis of the Role of African Philosophy and Ethics in Environment Management'. *Journal of Human Ecology* 53(2): 157–66.

Chomsky, N. 2007. 'Of Minds and Language'. *Biolinguistics* 1: 9–27.

Chua, K., W. Bellini, P. Rota et al. 2000. 'Nipah Virus: A Recently Emergent Deadly Paramyxovirus'. *Science* 288(5470): 1432–35.

Cissewski, J. and L. Luncz. 2021. 'Symbolic Signal Use in Wild Chimpanzee Gestural Communication? A Theoretical Framework'. *Frontiers in Evolutionary Psychology* 12: 718414

Cobley, P. 2016. *Cultural Implications of Biosemiotics*. New York: Springer.

Cobley, P., J. Deely, K. Kull and S. Petrilli, eds. 2011. *Semiotics Continues to Astonish*. Berlin: De Gruyter.

Cocroft, R. B., M. Gogala, P.S.M. Hilland and A. Wessel, eds. 2014. *Studying Vibrational Communication*. New York: Springer.

Colchester, M. 2004. 'Conservation Policy and Indigenous Peoples'. *Environmental Science and Policy* 7(3): 145–53.

Comyn, M. 1922. 'My Recollections of Karl Marx'. *The Nineteenth Century and After*, no. 91 (January), 161–69.

Cornelli, G., R. McKirahan and C. Macris, eds. 2013. *On Pythagoreanism*. Berlin: De Gruyter.

Cortopassi-Laurino, M., V. L. Imperatriz-Fonseca, D. W. Roubik et al. 2006. 'Global Meliponiculture: Challenges and Opportunities'. *Apidologie* 37(2): 275–92.

Coward, R. T., K. L. DeWeaver, F. E. Schmidt and R. W. Jackson. 1983. 'Distinctive Features of Rural Environments: A Frame of Reference for Mental Health Practice'. *International Journal of Mental Health* 12(1–2): 3–24.
Crist, E. 1999. *Images of Animals*. Philadelphia: Temple University Press.
Crist, E. 2004. 'Can an Insect Speak? The Case of the Honeybee Dance Language'. *Social Studies of Science* 34(1): 7–43.
Crist, E. 2019. *Abundant Earth: Toward an Ecological Civilisation*. Chicago: University of Chicago Press.
Cros, M. 2020. 'La mémoire longue des chauves-souris du Burkina: "Histoire d'Ebola ou pas"'. *Anthropologica* 62: 35–47.
Curley, E., ed. and trans. 1994. *A Spinoza Reader: The Ethics and Other Works*. Princeton, NJ: Princeton University Press.
Danchin, E., S. Nobel, A. Pocheville et al. 2018. 'Cultural Flies: Conformist Social Learning in Fruitflies Predicts Long-Lasting Mate-Choice Traditions'. *Science* 362(6418): 1025–30.
Dashper, K. 2016. 'Learning to Communicate: The Triad of (Mis)Communication in Horse Riding Lessons'. In *The Meaning of Horses: Biosocial Encounters*, edited by D. Davis and A. Maurstad, 101–15. London: Routledge.
Davis, D. and A. Maurstad. 2016. *The Meaning of Horses: Biosocial Encounters*. London: Routledge.
De la Cadena, M. 2015. *Earth Beings: Ecologies of Practice Across Andean Worlds*. Durham, NC: Duke University Press.
DEFRA. 2024. *Policy Paper: The Nature Recovery Network*. UK Government: Department for Environment, Food and Rural Affairs. Available at https://www.gov.uk/government/publications/nature-recovery-network/nature-recovery-network (accessed 28 January 2025).
Degnen, C. 2009. 'On Vegetable Love: Gardening, Plants, and People in the North of England'. *Journal of the Royal Anthropological Institute* 15: 151–67.
Deleuze, G. 1986. *Cinema 1: The Movement-Image*. Minneapolis: University of Minnesota Press.
Deleuze, G. and F. Guattari. 1987. *A Thousand Plateaus*, translated by B. Massumi. Minneapolis: University of Minnesota Press.
Deloughrey, E. 2017. 'Submarine Futures of the Anthropocene'. *Comparative Literature* 69(1): 32–44.
Demographia. 2023. *Demographia World Urban Areas*, 19th edn (31 August). http://www.demographia.com/db-worldua.pdf (accessed 27 January 2025).
Descola, P. 2013. *Beyond Nature and Culture*. Chicago: University of Chicago Press.
Despret, V. 2004. 'The Body We Care For: Figures of Anthropo-Zoo-Genesis'. *Body and Society* 10(2–3): 111–34.
Despret, V. 2015. 'Who Made Clever Hans Stupid?'. *Angelaki* 20(2): 77–85.
Despret, V. and B. Buchanan. 2016. *What Would Animals Say if We Asked the Right Questions?*. Minneapolis: University of Minnesota Press.
Dobson, J., J. Birch, P. Brindley et al. 2021. 'The Magic of the Mundane: The Vulnerable Web of Connections Between Urban Nature and Wellbeing'. *Cities* 108: 102989.
Dong, S., T. Lin, J. C. Nieh and K. Tan. 2023. 'Social Signal Learning of the Waggle Dance in Honey Bees'. *Science* 379: 1015–18.

Douglas, M. 2002 [1966]. *Purity and Danger: An Analysis of Concepts of Pollution and Taboo.* London: Routledge.

Drakopulos, L. 2020. 'New Materialist Approaches to Fisheries: The Birth of "Bycatch"'. *Environment and Society* 11(1): 100–114.

Dryzek, J. S. and J. Pickering. 2018. *The Politics of the Anthropocene.* Oxford: Oxford University Press.

Du Chaillu, P. B. 1900. *The World of the Great Forest: How Animals, Birds, Reptiles, Insects, Talk, Think, Work, and Live.* New York: Charles Scribner's Sons.

Eagleton, T. 1991. *Ideology: An Introduction.* London: Verso.

Elias, D. O., W. P. Maddison, C. Peckmezian, M. B. Girard and A. C. Mason. 2012. 'Orchestrating the Score: Complex Multimodal Courtship in the *Habronattus coecatus* Group of *Habronattus* Jumping Spiders (Araneae: Salticidae)'. *Biological Journal of the Linnean Society* 105: 522–47.

Elton, S. 2021. 'Growing Methods: Developing a Methodology for Identifying Plant Agency and Vegetal Politics in the City'. *Environmental Humanities* 13(1): 93–112.

Emmeche, C. and K. Kull, eds. 2011. *Towards a Semiotic Biology: Life is the Action of Signs.* London: Imperial College Press.

Engesser, S., A. R. Ridley and S. W. Townsend. 2016. 'Meaningful Call Combinations and Compositional Processing in the Southern Pied Babbler'. *PNAS* 113(21): 5976–81.

Etieyibo, E. 2017. 'Ubuntu and the Environment'. In *The Palgrave Handbook of African Philosophy*, edited by A. Afolayan and T. Falola, 633–57. New York: Palgrave Macmillan.

Evans, E. P. 1906. *The Criminal Prosecution and Capital Punishment of Animals.* London: Heinemann.

FAO. 2020. *Global Forest Resources Assessment 2020: Terms and Definitions.* Rome: Food and Agriculture Organization.

Fairhead, J. 2018. 'Technology, Inclusivity and the Rogue: Bats and the War Against the "Invisible Enemy"'. *Conservation and Society* 16(2): 170–80.

Fairhead, J. and M. Leach. 1995. 'False Forest History, Complicit Social Analysis: Rethinking Some West African Environmental Narratives'. *World Development* 23(6): 1023–36.

Fairhead, J. and M. Leach. 1996. *Misreading the African Landscape: Society and Ecology in a Forest-Savanna Mosaic.* Cambridge: Cambridge University Press.

Fairhead, J. and M. Leach. 1998. *Reframing Deforestation: Global Analyses and Local Realities—Studies in West Africa.* London: Routledge.

Fairhead, J. and M. Leach. 2003a. 'Termites, Society and Ecology: Perspectives from West Africa'. In *Insects in Oral Literature and Traditions*, edited by E. Motte-Florac and J.M.C. Thomas, 97–109. Paris-Louvain: Peeters-SELAF.

Fairhead, J. and M. Leach. 2003b. *Science, Society and Power.* Cambridge: Cambridge University Press.

Fairhead, J., M. Leach and D. Millimouno. 2021. 'Spillover or Endemic? Reconsidering the Origins of Ebola Virus Disease Outbreaks by Revisiting Local Accounts in Light of New Evidence from Guinea'. *BMJ Global Health* 6(4): e005783.

Fairhead, J., M. Leach and I. Scoones. 2012. 'Green Grabbing: A New Appropriation of Nature?'. *Journal of Peasant Studies* 39(2): 285–307.

Fairhead, J. and D. Millimouno. 2017. 'Ebola in Meliandou: Tropes of "Sustainability" at Ground Zero'. In *The Anthropology of Sustainability: Beyond Development and Progress*, edited by M. Brightman and J. Lewis, 165–81. Dordrecht: Springer.

Farías, I. and A. Blok, eds. 2016. *Urban Cosmopolitics: Agencements, Assemblies, Atmospheres*. London: Routledge.

Favareau, D., ed. 2010. *Essential Readings in Biosemiotics: Anthology and Commentary*. Dordrecht: Springer.

Feeley-Harnik, G. 2021. 'Lewis Henry Morgan: American Beavers and Their Works'. *Ethnos* 86(1): 21–43.

Fernandez, A. A. and M. Knörnschild. 2020. 'Pup Directed Vocalizations of Adult Females and Males in a Vocal Learning Bat'. *Frontiers in Ecology and Evolution*, Section 'Behavioral and Evolutionary Ecology' 8: 265.

Ferrigno, S., S. J. Cheyette, S. T. Piantadosi and J. F. Cantlon 2020. 'Recursive Sequence Generation in Monkeys, Children, U.S. Adults, and Native Amazonians'. *Science Advances* 2020(6): eaaz1002.

Figueiredo, A. F., B. Jens and G. Guggenberger. 2021. 'A Review of the Theories and Mechanisms Behind Underground Interactions'. *Frontiers in Fungal Biology* 2021(2): 735299.

Fijn, N. 2021. 'Human–Horse Sensory Engagement Through Horse Archery'. *The Australian Journal of Anthropology* 32: 58–79.

Fijn, N. and M. Kavesh. 2024. *Nurturing Alternative Futures: Living with Diversity in a More-than-Human World*. London: Routledge.

Filippi, P. 2016. 'Emotional and Interactional Prosody Across Animal Communication Systems: A Comparative Approach to the Emergence of Language'. *Frontiers in Psychology* 7 (art. 1393): 1–19.

Filippi, P., J. V. Congdon, J. Hoang et al. 2017. 'Humans Recognize Emotional Arousal in Vocalizations Across All Classes of Terrestrial Vertebrates: Evidence for Acoustic Universals'. *Proceedings of the Royal Society B: Biological Sciences* 284(1959): 1–9.

Fitzgerald, D., N. Manning, N. Rose and H. Fu. 2019. 'Mental Health, Migration and the Megacity'. *International Health* 11: S1–S6.

Fitz-Henry, E. 2022. 'Multispecies Justice: A View from the Rights of Nature Movement'. *Environmental Politics* 31(2): 338–59.

Flower, T. P., M. Gribble and A. R. Ridley. 2014. 'Deception by Flexible Alarm Mimicry in an African Bird'. *Science* 344(6183): 513–16.

Ford, A. 2019. 'Sport Horse Leisure and the Phenomenology of Interspecies Embodiment' *Leisure Studies* 38(3): 329–40.

Ford, N. and D. Brown. 2006. *Surfing and Social Theory: Experience, Embodiment and Narrative of the Dream Glide*. London: Routledge.

Foucault, M. 1982. *Power/Knowledge: Selected Interviews and Other Writings 1972–1977*, edited by C. Gordon. New York: Pantheon.

Frankal, C. 2020. 'The Stories We Tell Ourselves'. *Sea Change Project*. https://seachangeproject.com/stories/the-stories-we-tell-ourselves/ (accessed 27 January 2025).

Fraser, N. 2013. 'A Triple Movement?'. *New Left Review*, 81 (May–June), 119–32.

Fraser, J. A., A. Cosiaux, G. Walters et al. 2024. 'Defining the Anthropocene Tropical Forest: Moving Beyond "Disturbance" and "Landscape Domestication" with Concepts from African Worldviews'. *The Anthropocene Review* 11(3): 614–35.

Fraser, J. A., M. Leach and J. Fairhead. 2014. 'Anthropogenic Dark Earths in the Landscapes of Upper Guinea, West Africa: Intentional or Inevitable?'. *Annals of the Association of American Geographers* 104(6): 1222–38.

Frausin, V., J. A. Fraser, W. Narmah et al. 2014. '"God Made the Soil, but We Made It Fertile": Gender, Knowledge, and Practice in the Formation and Use of African Dark Earths in Liberia and Sierra Leone'. *Human Ecology* 42(5): 695–710.

Frisch, K. von. 1967. *The Dance Language and Orientation of Bees*. Cambridge, MA: Harvard University Press.

Full Fact. 2020. 'Fact Checks: How Many Birds Are Chickens'. 27 February. https://fullfact.org/environment/how-many-birds-are-chickens/ (accessed 26 January 2025).

Gabrić, P. 2022. 'Overlooked Evidence for Semantic Compositionality and Signal Reduction in Wild Chimpanzees (*Pan troglodytes*)'. *Animal Cognition* 25(3): 631–43.

Gagliano, M. 2018. *Thus Spoke the Plant: A Remarkable Journey of Groundbreaking Scientific Discoveries and Personal Encounters with Plants*. Berkeley, CA: North Atlantic Books.

Galef, B. G. and A. Whiten. 2017. 'The Comparative Psychology of Social Learning'. In *APA Handbook of Comparative Psychology*, edited by J. Call et al., 411–39. Washington, DC: American Psychological Association.

Game, A. 2001. 'Riding: Embodying the Centaur'. *Body and Society* 7(4): 1–12.

Geertz, C. 1973. *The Interpretation of Cultures*. Cambridge: Cambridge University Press.

Gentner, T. Q., K. M. Fenn, D. Margoliash and H. C. Nusbaum. 2006. 'Recursive Syntactic Pattern Learning by Songbirds'. *Nature* 440: 1204–7.

Ghosh, A. 2021. *The Nutmeg's Curse: Parables for a Planet in Crisis*. London: John Murray.

Gibson, D. 2018. 'Towards Plant-Centred Methodologies in Anthropology'. *Anthropology Southern Africa* 41(2): 92–103.

Gibson, D. and W. Ellis. 2018. 'Human and Plant Interfaces: Relationality, Knowledge and Practices'. *Anthropology Southern Africa* 41(2): 75–79.

Glaser, B. and J. Birk. 2012. 'State of the Scientific Knowledge on Properties and Genesis of Anthropogenic Dark Earths in Central Amazonia (Terra Preta de Índio)'. *Geochimica et Cosmochimica Acta* 82: 39–51.

Gould, J. and C. Gould. 1995 [1988]. *The Honey Bee*. New York: Scientific American Library.

Gramsci, A. 1971. *Selections from the Prison Notebooks of Antonio Gramsci*, edited and translated by Quintin Hoare and Geoffrey Nowell Smith. New York: International Publishers.

Green Connect. 2020. 'The 12 Principles of Permaculture'. https://green-connect.com.au/heres-your-guide-to-the-12-principles-of-permaculture/ (accessed 27 January 2025).

Greenough, P. 2003. 'Bio-Ironies of the Fractured Forest: India's Tiger Reserves'. In *In Search of the Rain Forest: New Ecologies for the Twenty-First Century*, edited by C. Slater, 167–203. Durham, NC: Duke University Press.

Grüter, C. and T. J. Czaczkes. 2019. 'Communication in Social Insects and How It Is Shaped by Individual Experience'. *Animal Behaviour* 151: 207–15.

Grüter, C. and E. Leadbeater. 2015. 'Insights from Insects about Adaptive Social Information Use'. *Trends in Ecology & Evolution* 29: 177–84.

Guddemi, P. 2020. *Gregory Bateson on Relational Communication: From Octopuses to Nations*. Berlin: Springer.

Habermas, J. 1984 [1981]. *Theory of Communicative Action, Volume One: Reason and the Rationalization of Society*, translated by T. A. McCarthy. Boston, MA: Beacon Press.

Halfwerk, W., J. Varkevisser, S. R. Ralph, E. Mendoza, C. Scharff and K. Riebel. 2019. 'Toward Testing for Multimodal Perception of Mating Signals'. *Frontiers in Ecology and Evolution* 2019(7): e124.

Haraway, D. 1991. *Simians, Cyborgs, and Women: The Reinvention of Nature*. New York: Routledge.

Haraway, D. 2003. *The Companion Species Manifesto: Dogs, People, and Significant Otherness*. Chicago: Prickly Paradigm Press.

Haraway, D. 2008. *When Species Meet*. Minneapolis: University of Minnesota Press.

Haraway, D. 2016. *Staying with the Trouble: Making Kin in the Chthulucene*. Durham, NC: Duke University Press.

Harcourt, W. and I. Nelson, 2015. *Practising Feminist Political Ecologies: Moving Beyond the 'Green Economy'*. The Hague: International Institute of Social Studies.

Harman, G. 2022. 'An Outline of Object-Oriented Philosophy'. *Science Progress* 96(2): 187–99.

Hartemink, A. E. 2016. 'The Definition of Soil since the Early 1800s'. In *Advances in Agronomy*, vol. 137, edited by D. L. Sparks, 73–126. New York: Academic Press.

Hasenjager, M. J., V. R. Franks and E. Leadbeater. 2022. 'From Dyads to Collectives: A Review of Honeybee Signalling'. *Behavioral Ecology and Sociobiology* 76: e124.

Hayward, E. 2010. 'Fingereyes: Impressions of Cup Corals'. *Cultural Anthropology* 25(4): 577–99.

Helmreich, S. 2009. *Alien Ocean: Anthropological Voyages in Microbial Seas*. Berkeley, CA: University of California Press.

Helmreich, S. 2011. 'Nature/Culture/Seawater'. *American Anthropologist*, n.s. 113(1): 132–44.

Helmreich, S. 2015. *Sounding the Limits of Life: Essays in the Anthropology of Biology and Beyond*. Princeton, NJ: Princeton University Press.

Helmreich, S. 2023. *A Book of Waves*. Durham, NC: Duke University Press.

Hersh, T. A., S. Gero, L. Rendell et al. 2022. 'Evidence from Sperm Whale Clans of Symbolic Marking in Non-Human Cultures'. *PNAS* 119(37): e2201692119.

Higham, J. P. and E. A. Hebets. 2013. 'An Introduction to Multimodal Communication'. *Behavioral Ecology and Sociobiology* 67: 1381–88.

Hinchliffe, S. and S. Whatmore. 2006. 'Living Cities: Towards a Politics of Conviviality'. *Science as Culture* 15(2): 123–38.

HLPE. 2019. *Agroecological and Other Innovative Approaches for Sustainable Agriculture and Food Systems that Enhance Food Security and Nutrition* (Report 14 of the High Level Panel of Experts on Food Security and Nutrition of the Committee on World Food Security). Rome: HLPE Secretariat. Available at https://openknowledge.fao.org/server/api/core/bitstreams/ff385e60-0693-40fe-9a6b-79bbef05202c/content (accessed 26 January 2025).

Hoeschele, M., D. C. Mann and B. Wagner. 2023. 'Using Knowledge About Human Vocal Behavior to Understand Acoustic Communication in Animals and the Evolution of Language and Music'. In *Acoustic Communication in Animals*, edited by Y. Seki, 1–25. Singapore: Springer.

Hoeschele, M., B. Wagner and D. C. Mann. 2023. 'Lessons Learned in Animal Acoustic Cognition Through Comparisons with Humans'. *Animal Cognition* 26: 97–116.

Hoffmeyer, J. 2008. *Biosemiotics: An Examination of the Signs of Life and the Life of Signs*. Scranton, PA: University of Scranton Press.

Hölldobler, B. and E. O. Wilson, 2009. *The Superorganism: The Beauty, Elegance, and Strangeness of Insect Societies*. New York: W. W. Norton.

Hutton, C. M. 1999. *Linguistics and the Third Reich: Mother-Tongue Fascism, Race and the Science of Language*. London: Routledge.

Hyde, W. W. 1916. 'The Prosecution and Punishment of Animals and Lifeless Things in the Middle Ages and Modern Times'. *University of Pennsylvania Law Review and American Law Register* 64(7): 696–730.

Iamblichus. 1818. *Iamblichus' Life of Pythagoras, or Pythagoric Life*, translated by Thomas Taylor. London: A. Valpy.

Ingold, T. 2000. *The Perception of the Environment: Essays on Livelihood, Dwelling and Skill*. London: Routledge.

Ingold, T. 2011. *Being Alive: Essays on Movement, Knowledge and Description*. New York: Routledge.

Ingold, T. and G. Pálsson, eds. 2013. *Biosocial Becomings: Integrating Social and Biological Anthropology*. Cambridge: Cambridge University Press.

Irvine, R. 2014. 'Thinking with Horses: Troubles with Subjects, Objects and Diverse Entities in Eastern Mongolia'. *Humanimalia* 6(1): 62–94.

Jakobson, R. 1965. 'Quest for the Essence of Language'. *International Council for Philosophy and Human Sciences* 13(51): 21–37

Jeffreys, M.D.Q. 1944. 'African Pterodactyls'. *Journal of the Royal African Society* 43(171): 72–74.

Jepson, P. 2019. 'Recoverable Earth: A Twenty-First Century Environmental Narrative'. *Ambio* 48: 123–30.

Johansen, J. D. 1988. 'The Distinction Between Icon, Index, and Symbol in the Study of Literature'. In *Semiotic Theory and Practice: Proceedings of the Third International Congress of the International Association for Semiotic Studies, Palermo, 1984, Volumes 1 + 2*, edited by M. Herzfeld and L. Melazzo, 497–504. Berlin: De Gruyter Mouton, 1988.

Johansen, J. D. 2002. *Literary Discourse: A Semiotic-Pragmatic Approach to Literature*. Toronto: University of Toronto Press.

Johnson, C. K., P. L. Hitchens, P. S. Pandit et al. 2020. 'Global Shifts in Mammalian Population Trends Reveal Key Predictors of Virus Spillover Risk'. *Proceedings of the Royal Society B* 287: 20192736.

Johnson, J. C. 1996. 'Maritime Anthropology'. In *The Encyclopedia of Cultural Anthropology*, edited by D. Levinson and M. Ember, vol. 3, 726–28. New York: Henry Holt.

Johnson, S. K., M. A. Fitza, D. A. Lerner et al. 2018. 'Risky Business: Linking *Toxoplasma gondii* Infection and Entrepreneurship Behaviours Across Individuals and Countries'. *Proceedings of the Royal Society B* 285: 20180822.

Jolly, A. 1966. 'Lemur Social Behavior and Primate Intelligence'. *Science* 153: 501–6.

Jones McVey, R. 2015. *Globetrotting: A Travelogue Exploring Horsemanship in Far-Flung Places*. London: J. A. Allen.

Jones McVey, R. 2022. 'Seeking Contact: British Horsemanship and Stances Toward Knowing and Being Known by (Animal) Others'. *Ethos* 50(4): 465–79.

Jones McVey, R. 2023. *Human-Horse Relations and the Ethics of Knowing*. London: Routledge.

Kamins, A., M. Rowcliffe and O. Restif. 2014. 'Ebola: Bats Get a Bad Rap When It Comes to Spreading Disease'. *The Conversation*. Available at http://theconversation.com/ebola-bats-get-a-bad-rap-when-it-comes-to-spreading-diseases-32785 (accessed 26 January 2025.

Karban, R. 2015. *Plant Sensing and Communication*. Berkeley, CA: University of California Press.

Karban, R. 2021. 'Plant Communication'. *Annual Review of Ecology, Evolution, and Systematics* 52(1): 1–24.

Karban, R., P. Grof-Tisza and C. Couchoux. 2022. 'Consistent Individual Variation in Plant Communication: Do Plants Have Personalities?'. *Oecologia* 199: 129–37.

Karban, R., L. H. Yang and K. F. Edwards. 2013. 'Volatile Communication Between Plants that Affects Herbivory: A Meta-Analysis'. *Ecology Letters* 17: 44–52.

Karlsson, F. 2012. 'Anthropomorphism and Mechanomorphism'. *Humanimalia* 3(2): 107–22.

Karst, J., M. D. Jones and J. D. Hoeksema. 2023. 'Positive Citation Bias and Overinterpreted Results Lead to Misinformation on Common Mycorrhizal Networks in Forests'. *Nature Ecology and Evolution* 7: 501–11.

Kavesh, M. A. 2021. 'The Flight of the Self: Exploring More-than-Human Companionship in Rural Pakistan'. *Australian Journal of Anthropology* 32: 42–57.

Keane, B. (2011). 'Traditional Māori religion—*ngā karakia a te Māori*: Spiritual Concepts'. In *Te Ara: The Encyclopedia of New Zealand*. https://teara.govt.nz/en/traditional-maori-religion-nga-karakia-a-te-maori/page-3 (accessed 27 January2025).

Keesing, R. M. 1974. 'Transformational Linguistics and Structural Anthropology'. *Cultural Hermeneutics* 2: 243–66.

Kennedy, C., J. Cuddihy and J. Engel-Yan. 2007. 'The Changing Metabolism of Cities'. *Journal of Industrial Ecology* 11(2): 43–59.

Kennedy, J. S. 1992. *The New Anthropomorphism*. Cambridge: Cambridge University Press.

Kerby, L. 2010. 'Surfing and Spirituality'. *Social Sciences Capstone Projects*, Paper 6. Available at https://core.ac.uk/download/pdf/48854033.pdf (accessed 27 January 2025.

Kimmerer, R. W. 2015. *Braiding Sweetgrass: Indigenous Wisdom, Scientific Knowledge and the Teachings of Plants*. Minneapolis: Milkweed Editions.

Kirksey, E., ed. 2014. *The Multispecies Salon*. Durham, NC: Duke University Press.

Knörnschild, M. and A. Fernandez. 2020. 'Do Bats Have the Necessary Prerequisites for Symbolic Communication?'. *Frontiers in Psychology* 11: 571678.

Kohn, E. 2013. *How Forests Think: Towards an Anthropology Beyond the Human*. Berkeley, CA: University of California Press.

Kockelman, P. 2011. 'A Mayan Ontology of Poultry: Selfhood, Affect, Animals, and Ethnography'. *Language in Society* 40(4): 427–54.

Kopnina, H. 2016. 'Half the Earth for People (or More)? Addressing Ethical Questions in Conservation'. *Biological Conservation* 203: 176–85.

Kothari, A., A. Salleh, A. Escobar, F. Demaria and A. Acosta, eds. 2019. *Pluriverse: A Post-Development Dictionary*. New Delhi: Tulika Books in association with AuthorsUpFront.

Krzywoszynska, A. 2019. 'Caring for Soil Life in the Anthropocene: The Role of Attentiveness in More-Than-Human Ethics'. *Transactions of the Institute of British Geographers* 44(4): 661–75.

Krzywoszynska, A. 2020. 'Nonhuman Labor and the Making of Resources: Making Soils a Resource through Microbial Labor'. *Environmental Humanities* 12(1): 227–49.

Krzywoszynska, A., S. Banwart, and D. Blacker. 2020. 'To Know, to Dwell, to Care: Towards an Actionable, Place-Based Knowledge of Soils'. In *Thinking with Soil: Material Politics and Social Theory*, edited by F. Salazar, C. Granjou, M. Kearnes, A. Krzywoszynska and M. Tironi. London: Bloomsbury.

Krzywoszynska, A. and G. Marchesi. 2020. 'Toward a Relational Materiality of Soils: Introduction'. *Environmental Humanities* 12(1): 190–204.

Kull, K. 1998. 'On Semiosis, Umwelt, and Semiosphere'. *Semiotica* 120(3/4): 299–310.

Kull, K. 1999. 'Biosemiotics in the Twentieth Century: A View from Biology'. *Semiotica* 127(1/4): 385–414.

Kull, K. 2003. 'Thomas A. Sebeok and Biology: Building Biosemiotics'. *Cybernetics and Human Knowing* 10: 47–60.

Kull, K. 2010. 'Umwelt'. In *The Routledge Companion to Semiotics*, edited by P. Cobley, 348–49. London: Routledge.

Kull, K., T. W. Deacon, C. Emmeche, J. Hoffmeyer and F. Stjernfelt. 2009. 'Theses on Biosemiotics: Prolegomena to a Theoretical Biology'. *Biological Theory* 4: 167–73.

Lakoff, G. and M. Johnson. 1980. *Metaphors We Live By*. Chicago: University of Chicago Press.

Laland, K. N., N. Atton and M. M. Webster. 2011. 'From Fish to Fashion: Experimental and Theoretical Insights into the Evolution of Culture'. *Philosophical Transactions of the Royal Society B* 366: 958–68.

Lamara, F., M. Calmet and S. Hayes, eds. 2024. *The Rights of Nature*. Paris: Agence France de Développement.

Lameira, A. R., M. E. Hardus, A. Ravignani, T. Raimondi and M. Gamba. 2023. 'Recursive Self-Embedded Vocal Motifs in Wild Orangutans'. *eLife* 12: RP88348.

Lasco, G. 2020. 'How COVID-19 Is Changing People's Relationships with Houseplants'. *Sapiens*, 17 September. https://www.sapiens.org/column/entanglements/covid-19-houseplants (accessed 26 January 2025).

Latour, B. 1987. *Science in Action: How to Follow Scientists and Engineers Through Society*. Milton Keynes: Open University Press.

Latour, B. 1999. *Pandora's Hope*. Cambridge, MA: Harvard University Press.

Latour, B. 2004. *Politics of Nature: How to Bring the Sciences into Democracy*, translated by C. Porter. Cambridge, MA: Harvard University Press.

Latour, B. 2005. *Reassembling the Social: An Introduction to Actor-Network-Theory*. Oxford: Oxford University Press.

Latour, B. 2018. *Down to Earth: Politics in the New Climatic Regime*. New York: Wiley.

Leach, M. 1994. *Rainforest Relations: Gender and Resource Use Among the Mende of Gola, Sierra Leone*. Edinburgh: Edinburgh University Press.

Leach, M. 2000. 'New Shapes to Shift: War, Parks and the Hunting Persona in Modern West Africa'. *Journal of the Royal Anthropological Institute* 6(4): 577–95.

Leach, M., B. Bett, M. Said et al. 2017. 'Local Disease–Ecosystem–Livelihood Dynamics: Reflections from Comparative Case Studies in Africa'. *Philosophical Transactions of the Royal Society B* 372: 1725.

Leach, M., H. MacGregor, S. Ripoll, I. Scoones and A. Wilkinson. 2022. 'Rethinking Disease Preparedness: Incertitude and the Politics of Knowledge'. *Critical Public Health* 32(1): 82–96.

Leach, M. and I. Scoones. 2015a. 'Mobilizing for Green Transformations'. In *The Politics of Green Transformations*, edited by I. Scoones, M. Leach and P. Newell, 119–33. London: Routledge.

Leach, M. and I. Scoones, eds. 2015b. *Carbon Conflicts and Forest Landscapes in Africa*. London: Routledge.

Leach, M., I. Scoones and A. Stirling, 2010. *Dynamic Sustainabilities: Technology, Environment, Social Justice*. London: Routledge.

Leach, M., I. Scoones and B. Wynne, eds. 2005. *Science and Citizens: Globalisation and the Challenge of Engagement*. London: Routledge.

Leadbeater, E. and L. Chittka. 2007. 'Social Learning in Insects—From Miniature Brains to Consensus Building'. *Current Biology* 17: R703–R713.

Leeson, P. T. 2013. 'Vermin Trials'. *The Journal of Law and Economics* 56(3): 811–36.

Lefebvre, H. and 1991 [1974]. *The Production of Space*, translated by D. Nicholson-Smith. Oxford: Blackwell.

Lefèvre, C., F. Rekik, V. Alcantara and L. Wiese. 2017. *Soil Organic Carbon: The Hidden Potential*. Rome: Food and Agriculture Organization of the United Nations.

Lenin, V. I. 2008. *Revolution, Democracy, Socialism*. London: Pluto Press.

Leroux, M., A. B. Bosshard, B. Chandia, A. Manser, K. Zuberbühler and S. W. Townsend. 2021. 'Chimpanzees Combine Pant Hoots with Food Calls into Larger Structures'. *Animal Behaviour* 179: 41–50.

Leroux, M. and S. W. Townsend. 2020. 'Call Combinations in Great Apes and the Evolution of Syntax'. *Animal Behaviour: Cognitive* 7(2): 131–39.

Leroy, E., B. Kumulungui, X. Pourrut et al. 2005. 'Fruit Bats as Reservoirs of Ebola Virus'. *Nature* 438(7068): 575–76.

Lestel, D. 1998. 'L'innovation cognitive dans des communautés hybrids homme/animal de partage de sens, d'intérêts et d'affects'. *Intellectica*, 1/2(26/27), 203–26.

Lestel, D. 2002. 'The Biosemiotics and Phylogenesis of Culture'. *Social Science Information* 41(1): 35–68.

Lestel, D. and H. Taylor. 2013. 'Shared Life: An Introduction'. *Social Science Information* 52(2): 183–86.

Levin, S. A. 2005. 'Self-Organization and the Emergence of Complexity in Ecological Systems'. *BioScience* 55(12): 1075–79.

Lévi-Strauss, C. 1963 [1958]. *Structural Anthropology*, translated by C. Jacobson and B. G. Schoepf. London: Allen Lane.

Liao, D. A., K. F. Brecht, M. Johnson and A. Nieder. 2022. 'Recursive Sequence Generation in Crows'. *Scientific Advances* 8: eabq3356.

Lindauer, M. 1971 [1961]. *Communication Among Social Bees*. Cambridge, MA: Harvard University Press.

Lisdorf, A. 2021. 'How Viewing Cities as Organisms Can Change the Way We Use and Consume Resources'. *Shareable*, 2 September. https://www.shareable.net/how-viewing-cities-as-organisms-can-change-the-way-we-use-and-consume-resources (accessed 27 January 2025).

Lorimer, J. 2020. *The Probiotic Planet: Using Life to Manage Life*. Minneapolis: University of Minnesota Press.

Lorimer, J. and T. Hodgetts. 2024. *More-than-Human*. London: Routledge.

Luby, S., M. Rahman, J. Hossain et al. 2006. 'Foodborne Transmission of Nipah Virus, Bangladesh'. *Emerging Infectious Diseases* 12(12): 1888–94.

Mac Aodha, O., R. Gibb, K. E. Barlow et al. 2018. 'Bat Detective: Deep Learning Tools for Bat Acoustic Signal Detection'. *PLOS Computational Biology* 14(3): e1005995.

Macy, J. and M. Y. Brown. 2014. *Coming Back to Life: The Updated Guide to the Work That Reconnects*. Gabriola Island (British Columbia): New Society Publishers.

Mancuso, S. 2017. *The Revolutionary Genius of Plants: A New Understanding of Plant Intelligence and Behavior*. New York: Atria Books.

Mann, D. C., W. T. Fitch, H.-W. Tu and M. Hoeschele. 2021. 'Universal Principles Underlying Segmental Structures in Parrot Song and Human Speech'. *Scientific Reports* 11 (776): 1–13.

Mann, D. C. and M. Hoeschele. 2020. 'Segmental Units in Nonhuman Animal Vocalization as a Window into Meaning, Structure, and the Evolution of Language'. *Animal Behaviour and Cognition* 7(2): 151–58.

Marchesi, G. 2020. 'Justus von Liebig Makes the World: Soil Properties and Social Change in the Nineteenth Century'. *Environmental Humanities* 12(1): 205–26.

Marder, M. 2013a. 'What Is Plant-Thinking?'. *Klēsis—revue philosophique* 25, Philosophies de la nature, 124–43.

Marder, M. 2013b. *Plant Thinking: A Philosophy of Vegetal Life*. New York: Columbia University Press.

Marino, L. 2017. 'Thinking Chickens: A Review of Cognition, Emotion, and Behavior in the Domestic Chicken'. *Animal Cognition* 20(2): 127–47.

Markowitz, J. E., E. Ivie, L. Kligler and T. J. Gardner. 2013. 'Long-Range Order in Canary Song'. *PLOS Computational Biology* 2013(9): e1003052.

Marler, P. and M. Tamura. 1964. 'Song "Dialects" in Three Populations of White-Crowned Sparrows'. *Science* 146: 1483–86.

Marris, E. 2013. *Rambunctious Garden: Saving Nature in a Post-Wild World*. New York: Bloomsbury.

Marya, R. and R. Patel. 2021. *Inflamed: Deep Medicine and the Anatomy of Injustice*. New York: Allen Lane.

Maurstad, A., D. L. Davis and S. Cowles. 2013. 'Co-Being and Intra-Action in Horse-Human Relationships: A Multispecies Ethnography of Be(com)ing Human and Be(com)ing Horse'. *Social Anthropology* 21(3): 322–35.

Maynard-Smith, J. and D. Harper. 2003. *Animal Signals*. Oxford: Oxford University Press.

Mbembe, A. 2019. *Necropolitics*. Durham, NC: Duke University Press.

McAfee, K. 1999. 'Selling Nature to Save It? Biodiversity and Green Developmentalism'. *Environment and Planning D: Society and Space* 17(2): 133–54

McKenzie, P. 2022. 'How Maori Stepped In to Save a Towering Tree Crucial to Their Identity'. *The New York Times*, 8 March.

McLauchlan, L. 2021. 'Multispecies Ethnography'. In *Handbook of Historical Animal Studies*, edited by M. Roscher, A. Krebber and B. Mizelle, 393–408. Berlin: De Gruyter Oldenbourg.

Meijer, E. 2019. *When Animals Speak: Toward an Interspecies Democracy*. New York: New York University Press.

Merchant, C. 1990. *The Death of Nature: Women, Ecology and the Scientific Revolution*. San Francisco: Harper.

Midgley, M. 1983. *Animals and Why They Matter*. Athens, GA: University of Georgia Press.

Mies, M. and V. Shiva. 1993. *Ecofeminism*. Halifax, Nova Scotia: Fernwood Publications.

Miller, T. L. 2019. *Plant Kin: A Multispecies Ethnography in Indigenous Brazil*. Austin, TX: University of Texas Press.

Minasny, B., B. P. Malone, A. B. McBratney et al. 2017. 'Soil Carbon 4 per Mille'. *Geoderma* 292: 59–86.

Monbiot, G. 2023. *Regenesis: Feeding the World Without Devouring the Planet*. London: Penguin.

Monchanin, C., A. Blanc-Brude, E. Drujont et al. 2021. 'Chronic Exposure to Trace Lead Impairs Honey Bee Learning'. *Ecotoxicology and Environmental Safety* 212: 112008.

Montaigne, M. de. 1991 [1580]. 'An Apology for Raymond Sebond'. In Montaigne, *The Complete Essays*, edited and translated by M. A. Screech, Bk 2, ch. 12. London: Allen Lane.

Moore, A. and J. Jacka. 2020. 'Introduction: Oceans'. *Environment and Society* 11(1): 1–4.

Moore, L. J. and M. Kosut. 2013. 'Among the Colony: Ethnographic Fieldwork, Urban Bees and Intraspecies Mindfulness'. *Ethnography*, 15(4): 516–39.

Moore-Colyer, R. 2004. 'Towards "Mother Earth": Jorian Jenks, Organicism, the Right and the British Union of Fascists'. *Journal of Contemporary History* 39(3): 353–71.

Morell, V. (2023). 'Dolphins and Humans Team Up to Catch Fish in Brazil: Marine Mammals May Even Be Training Their Human Counterparts'. *Science Adviser*, 30 January. https://www.science.org/content/article/dolphins-and-humans-team-catch-fish-brazil (accessed 28 January 2025).

Morgan, C. L. 1903. 'Other Minds than Ours'. In *An Introduction to Comparative Psychology*, edited by W. Scott. London: Walter Scott.

Morgan, L. H. 2018 [1868]. *The American Beaver and His Works*. Abingdon: Franklin Classics.

Moser, K. 2022. 'Dominique Lestel's Pioneering Biosemiotic Vision of "The Enchanted Space of Trans-Specific Communication"'. In *Contemporary French Environmental Thought in the Post-COVID-19 Era*, edited by K. Moser, 189–226. Dordrecht: Springer.

Murray, T. G., J. Zeil and R. D. Magrath. 2017. 'Sounds of Modified Flight Feathers Reliably Signal Danger in a Pigeon'. *Current Biology* 27: 3520–25.

Mustill, T. 2022. *How to Speak Whale: A Voyage into the Future of Animal Communication*. London: William Collins.

Nadasdy, P. 2007. 'The Gift in the Animal: The Ontology of Hunting and Human–Animal Sociality'. *American Ethnologist* 34: 25–43.

Nadkarni, N. 2008. *Between Earth and Sky: Our Intimate Connections to Trees*. Berkeley, CA: University of California Press.

Nagel, T. 1974. 'What Is It Like to Be a Bat?'. *The Philosophical Review* 83(4): 435–50.

Nahar, N., U. K. Mondal, M. J. Hossain et al. 2014. 'Piloting the Promotion of Bamboo Skirt Barriers to Prevent Nipah Virus Transmission Through Date Palm Sap in Bangladesh'. *Global Health Promotion* 21(4): 7–15.

Nathen, T. 2018. '"Being Attentive": Exploring Other-than-Human Agency in Medicinal Plants Through Everyday Rastafari Plant Practices'. *Anthropology Southern Africa* 41(2): 115–26.

Neer, R. M. 2013. *Napalm: An American Biography*. Cambridge, MA: Harvard University Press.

Nelson, R. 1983. *Make Prayers to the Raven: A Kokuyon View of the Northern Forest*. Chicago: University of Chicago Press.

Nhemachena, A. 2016. 'Animism, Coloniality and Humanism: Reversing the Empire's Framing of Africa'. In *Theory, Knowledge, Development and Politics: What Role for the Academy in the*

Sustainability of Africa?, edited by M. Mawere and A. Nhemachena, 13–54. Bamenda: Langaa Research and Publishing Common Initiative Group.

Norberg, J. 2004. 'Biodiversity and Ecosystem Functioning: A Complex Adaptive Systems Approach'. *Limnology and Oceanography* 49(4): 1269–77.

Novellino, D. 2019. 'The Relevance of Myths and Worldviews in Pälawan Classification, Perceptions, and Management of Honey Bees'. In *Ethnobiology and Biocultural Diversity: Proceedings of the Seventh International Congress of Ethnobiology*, edited by J. R. Stepp, F. S. Wyndham and R. K Zarger, 189–206. Athens, GA: University of Georgia, The International Society of Ethnobiology.

Nutsford, D., A. Pearson and S. Kingham. 2013. 'An Ecological Study Investigating the Association Between Access to Urban Green Space and Mental Health'. *Public Health* 127(11): 1005–11.

Nyamnjoh, F. B. 2017. *Drinking From the Cosmic Gourd: How Amos Tutuola Can Change Our Minds*. Bamenda: Langaa Research & Publishing CIG.

Nyamnjoh, F. B. 2020. *Decolonising the Academy: A Case for Convivial Scholarship*. Carl Schlettwein Lectures 14. Basel: Basler Afrika Bibliographien.

O'Brien, A. T. 2020. 'Ethical Acknowledgment of Soil Ecosystem Integrity amid Agricultural Production in Australia'. *Environmental Humanities* 12(1): 267–84.

O'Bryan, L., M. Beier and E. Salas. 2020. 'How Approaches to Animal Swarm Intelligence Can Improve the Study of Collective Intelligence in Human Teams'. *Journal of Intelligence* 8(1): 9.

Ogden, L. 2011. *Swamplife: People, Gators, and Mangroves Entangled in the Everglades*. Minneapolis: University of Minnesota Press.

Ogden, L., B. Hall and K. Tanita. 2013. 'Animals, Plants, People, and Things: A Review of Multispecies Ethnography'. *Environment and Society: Advances in Research* 4(1): 5–24.

Ohemeng, F., E. T. Lawson, J. Ayivor, M. Leach, L. Waldman and Y. Ntiamoa-Baidu. 2017. 'Socio-Cultural Determinants of Human–Bat Interaction in Rural Ghana'. *Anthrozoos* 30(2): 181–94.

Olival, K. 2016. 'To Cull or Not to Cull, Bat Is the Question'. *EcoHealth* 13(1): 6–8.

Oliveira, R. F. and R. Bshary. 2021. 'Expanding the Concept of Social Behavior to Interspecific Interactions'. *Ethology* 127(10): 758–73.

Olivero, J., J. E. Fa, R. Real et al. 2017. 'Recent Loss of Closed Forests Is Associated with Ebola Virus Disease Outbreaks'. *Scientific Reports* 2017(7): 14291.

Ouattara, K., A. Lemasson and K. Zuberbühler. 2009. 'Campbell's Monkeys Concatenate Vocalizations into Context-Specific Call Sequences'. *PNAS* 106(51): 22026–31.

Owens, P. R. and E. M. Rutledge. 2005. 'Morphology'. In *Encyclopedia of Soils in the Environment*, edited by D. Hillel, 511–20. Oxford: Elsevier Academic Press.

Pachirat, T. 2011. *Every Twelve Seconds: Industrialized Slaughter and the Politics of Sight*. New Haven, CT: Yale University Press.

Pálsson, G. 1998. 'The Birth of the Aquarium: The Political Ecology of Icelandic Fishing'. In *The Politics of Fishing*, edited by T. Gray, 209–27. London: Macmillan.

Patterson, J., K. Schulz, J. Vervoort, S. Van Der HelM. Sethi and A. Barau. 2017. 'Exploring the Governance and Politics of Transformations Towards Sustainability'. *Environmental Innovation and Societal Transitions* 24: 1–16.

Pauly, D. 1995. 'Anecdotes and the Shifting Baseline Syndrome of Fisheries'. *Trends in Ecology and Evolution* 10: art. 430.

Paykel, E., R. Abbott, R. Jenkins, T. Brugha and H. Meltzer. 2000. 'Urban–Rural Mental Health Differences in Great Britain: Findings from the National Morbidity Survey'. *Psychological Medicine* 30(2): 269–80.

Pearce, M. 2019. 'The Healing Power of Horses'. 1 May. https://touchedbyahorse.com/the-healing-power-of-horses/ (accessed 26 January 2025).

Pearce, T. 2010. '"A great complication of circumstances"—Darwin and the Economy of Nature'. *Journal of the History of Biology* 43(3): 493–528.

Peemot, V. S. 2017. 'We Eat Whom We Love: Hippophagy Among Tyvan Herders'. *Inner Asia* 19(1): 133–56.

Peen, J., R. A. Schoevers, A. T. Beekman and J. Dekker. 2010. 'The Current Status of Urban–Rural Differences in Psychiatric Disorders'. *Acta Psychiatrica Scandinavica* 121(2): 84–93.

Peirce, C. S. 1960 [193132]. *Collected Papers, Volumes I and II: Principles of Philosophy and Elements of Logic*, edited by C. Hartshorne and P. Weiss. Cambridge, MA: Harvard University Press.

Pepperberg, I. M. 2017. 'Symbolic Communication in Nonhuman Animals'. In *APA Handbook of Comparative Psychology: Basic Concepts, Methods, Neural Substrate, and Behavior*, edited by J. Call et al., 663–79. Washington, DC: American Psychological Association.

PerfectBee. n.d. 'The Language of Bees'. https://www.perfectbee.com/blog/the-language-of-bees (accessed 26 January 2025).

Peters, J., O. Lebrasseur, E. K. Irving-Pease and G. Larson. 2022. 'The Biocultural Origins and Dispersal of Domestic Chickens'. *PNAS* 119(24): e2121978119.

Petrović, G. 1991. 'Praxis'. In *The Dictionary of Marxist Thought*, 2nd edn, edited by T. Bottomore, L. Harris, V. G. Kiernan and R. Miliband, 435–40. Oxford: Blackwell.

Phillips, C. and E. Schulz. 2021. 'Greening Home: Caring for Plants Indoors'. *Australian Geographer* 52(4): 373–89.

Phillips, J. 2006. 'Agencement/Assemblage'. *Theory, Culture & Society* 23(2–3): 108–9.

Piqueret, B., J.-C. Sandoz and P. d'Ettorre. 2019. 'Ants Learn Fast and Do Not Forget: Associative Olfactory Learning, Memory and Extinction in *Formica fusca*'. *Royal Society Open Science* 2019(6): 190778.

Piva, H. C. 2023. 'Semiotically Mediated Human–Bee Communication in the Practice of Brazilian Meliponiculture'. *Biosemiotics* 16: 105–24.

Pleyer, M., R. Lepic, and S. Hartmann. 2022. 'Compositionality in Different Modalities: A View from Usage-Based Linguistics'. *International Journal of Primatology* 45(3): 670–702.

Plumwood, V. 2003. *Feminism and the Mastery of Nature*. London: Routledge.

Polanyi, K. 2001 [1944]. *The Great Transformation: The Political and Economic Origins of Our Time*. Boston, MA: Beacon Press.

Pony Club, The. 2009 [1950]. *The Manual of Horsemanship*. Amersham: Halstan & Co.

Pouw, W., S. Proksch, L. Drijvers et al. 2021. 'Multilevel Rhythms in Multimodal Communication'. *Philosophical Transactions of the Royal Society* 376: 20200334.

Powers, R. 2018. *The Overstory*. New York: W. W. Norton.

Prat, Y. 2019. 'Animals Have No Language, and Humans Are Animals Too'. *Perspectives on Psychological Science* 14: 885–93.

Prat, Y., L. Azoulay, R. Dor, and Y. Yovel. 2017. 'Crowd Vocal Learning Induces Vocal Dialects in Bats: Playback of Conspecifics Shapes Fundamental Frequency Usage by Pups'. *PLOS Biology* 15(10): e2002556.

Preiser, R., R. Biggs, A. De Vos and C. Folke. 2018. 'Social-Ecological Systems as Complex Adaptive Systems: Organizing Principles for Advancing Research Methods and Approaches'. *Ecology and Society* 23(4): e46

Price, A. 2018. 'The Living City vs. the Mechanical City'. *Strong Towns*, 2 May. https://www.strongtowns.org/journal/2018/5/1/the-living-city-vs-the-mechanical-city (accessed 27 January 2025).

Proulx, S. R., D.E.L. Promislow and P. C. Phillips 2005. 'Network Thinking in Ecology and Evolution'. *Trends in Ecology & Evolution* 20(6): 345–53.

Puig de la Bellacasa, M. 2015. 'Making Time for Soil: Technoscientific Futurity and the Pace of Care'. *Social Studies of Science* 45(5): 691–716.

Puig de la Bellacasa, M. 2017. *Matters of Care: Speculative Ethics in More-than-Human Worlds.* Minneapolis: University of Minnesota Press.

Purewal, R., R. Christley, K. Kordas et al. 2019. 'Socio-Demographic Factors Associated with Pet Ownership Amongst Adolescents from a UK Birth Cohort'. *BMC Veterinary Research* 15: art. 334.

Quick, N. J. and V. M. Janik. 2012. 'Bottlenose Dolphins Exchange Signature Whistles When Meeting at Sea'. *Proceedings of the Royal Society B* 279: 2539–45.

Raworth, K. 2017, *Doughnut Economics*. White River Junction, VT: Chelsea Green Publisher.

Rey, A. and J. Fagot. 2023. 'Associative Learning Accounts for Recursive-Structure Generation in Crows'. *Learning and Behavior* 51: 347–48.

Reznkova, Z. 2023. 'Information Theory Opens New Dimensions in Experimental Studies of Animal Behaviour and Communication'. *Animals* 13: e1174.

Richardson, M., J. Dobson, D. J. Abson et al. 2020. 'Applying the Pathways to Nature Connectedness at a Societal Scale: A Leverage Points Perspective'. *Ecosystems and People* 16(1): 387–401.

Rodríguez Higuera, C. 2020. 'Everything Seems So Settled Here: The Conceivability of Post-Peircean Biosemiotics'. *Sign Systems Studies* 47(3/4): 420–35.

Rosevear, D. R. 1979. 'Oban Revisited'. *The Nigerian Field* 44: 75–81.

Roys Farm. 2024. https://www.roysfarm.com/communicating-with-chickens/ (accessed 26 January 2025).

Rulli, M., M. Santini, D. Hayman and P. D'Odorico. 2017. 'The Nexus Between Forest Fragmentation in Africa and Ebola Virus Disease Outbreaks'. *Scientific Reports* 7: 41613.

Rydin, Y. 2013. 'Using Actor-Network Theory to Understand Planning Practice: Exploring Relationships Between Actants in Regulating Low-Carbon Commercial Development'. *Planning Theory* 12(1): 23–45.

Sabo, G. and D. Sabo, 1985. 'Belief Systems and the Ecology of Sea Mammal Hunting Among the Baffinland Eskimo'. *Arctic Anthropology* 22(2): 77–86

Safina, C. 2021. *Becoming Wild: How Animal Cultures Raise Families, Create Beauty, and Achieve Peace.* London: Picador.

Sagan, D. 2013. 'The Human Is More Than Human: Interspecies Communities and the New "Facts of Life"'. In *Cosmic Apprentice: Dispatches from the Edges of Science*, edited by D. Sagan, 17–32. Minneapolis: University of Minnesota Press, 2013

Salazar, J. F., C. Granjou, M. Kearnes, A. Krzywoszynska and M. Tironi, eds. 2020. *Thinking with Soils: Material Politics and Social Theory*. London: Bloomsbury.

SAS (2025). *Surfers Against Sewage*. https://www.sas.org.uk/ (accessed 27 January 2025).

Saussure, F. de. 2011 [1959]. *Course in General Linguistics*, edited by P. Meisel and H. Saussy and translated by W. Baskin. New York: Columbia University Press.

Schlenker, P., E. Chemla, A. M. Schel et al. 2016. 'Formal Monkey Linguistics'. *Theoretical Linguistics* 42: 1–90.

Schlosberg, D. 2004. 'Reconceiving Environmental Justice: Global Movements and Political Theories'. *Environmental Politics* 13(3) 517–40.

Schneeberger, K. and C. Voigt. 2016. 'Zoonotic Viruses and Conservation of Bats'. In *Bats in the Anthropocene: Conservation of Bats in a Changing World*, edited by C. Voigt and T. Kingston, 263–92. Berlin: Springer.

Scoones, I. 2015. 'Transforming Soils: Transdisciplinary Perspectives and Pathways to Sustainability'. *Current Opinion in Environmental Sustainability* 15: 20–24.

Scoones, I., M. Leach and A. Stirling, 2015. *The Politics of Green Transformations*. London: Routledge.

Scoones, I., A. Stirling, D. Abrol et al. 2020. 'Transformations to Sustainability: Combining Structural, Systemic and Enabling Approaches'. *Current Opinion in Environmental Sustainability* 42: 65–75.

Scott, E., D. E. Edgley, A. Smith et al. 2023. 'Lateral Line Morphology, Sensory Perception and Collective Behaviour in African Cichlid Fish'. *Royal Society Open Science* 10: 221478.

Sea Change Trust (2024). *Sea Change Project*. https://seachangeproject.com/ (accessed 27 January 2025).

Searcy, W. A., L. M. Chronister and S. Nowicki. 2023. 'Syntactic Rules Predict Song Type Matching in a Songbird'. *Behavioral Ecology Sociobiology* 77(12). http://www.doi.org/10.1007/s00265-022-03286-3

Searcy, W. A., J. Soha, S. Peters and S. Nowicki,. 2022. 'Long-Distance Dependencies in Birdsong Syntax'. *Proceedings of the Royal Society B* 289.

Sebeok, T. A. 1967. 'Aspects of Animal Communication: The Bees and Porpoises'. *Review of General Semantics* 24(1): 59–83.

Sebeok, T. A. 1972. *Perspectives in Zoosemiotics*. The Hague: Mouton.

Sebeok, T. A., ed. 1977. *How Animals Communicate*. Bloomington, IN: Indiana University Press.

Sebeok, T. A. 1996. 'Signs, Bridges, Origins'. In *Origins of Language*, edited by J. Trabant, 89–115. Budapest: Collegium.

Sebeok, T.A. 1997. 'Give Me Another Horse'. In *Reading Eco: An Anthology*, edited by R. Capozzi. 276–82. Bloomington, IN: Indiana University Press.

Sebeok, T. A., J. Hoffmeyer and C. Emmeche, eds. 1999. *Biosemiotica*. Berlin: Mouton de Gruyter.

Sebeok, T. A. and J. Umiker-Sebeok, eds. 1992. *Biosemiotics: The Semiotic Web 1991*. Berlin: Mouton de Gruyter.

Seeley, T. D. 2010. *Honeybee Democracy*. Princeton, NJ: Princeton University Press.

Senthurran, S. and A. C. Mason. 2021. 'Vibratory Communication in a Black Widow Spider (*Latrodectus hesperus*): Signal Structure and Signalling Mechanisms'. *Animal Behaviour* 174: 217–35.

Seyfarth, R. M., D. L. Cheney and P. Marler. 1980. 'Monkey Responses to Three Different Alarm Calls: Evidence of Predator Classification and Semantic Communication'. *Science* 210: 801–3.

Shanahan, D., R. Fuller, R. Bush, B. Lin and K. Gaston. 2015. 'The Health Benefits of Urban Nature: How Much Do We Need?'. *BioScience* 65(5): 476–85.

Sharpless, I. 2016. 'Politics and the Signs of Animal Life: Biosemiotics, Aristotle, and Human-Animal Relations'. In *Humans and Animals: Intersecting Lives and Worlds*, edited by A. Höing and A. M. Bennett, 25–39. Oxford: Inter-Disciplinary Press.

Sheldrake, M. 2020. *Entangled Life: How Fungi Make Our Worlds, Change Our Minds and Shape Our Futures*. London: Bodley Head.

Shiva, V. 2008. *Soil Not Oil: Environmental Justice in a Time of Climate Crisis*. Berkeley, CA: North Atlantic Books.

Sieradzki, A. and H. Mikkola. 2022. 'Bats in Folklore and Culture: A Review of Historical Perceptions Around the World'. In *Bats: Disease-Prone but Beneficial*, edited by H. Mikkola. IntechOpen. https://www.doi.org/10.5772/intechopen.102368

Simard, S. W. 2018. 'Mycorrhizal Networks Facilitate Tree Communication, Learning, and Memory'. In *Memory and Learning in Plants. Signaling and Communication in Plants*, edited by F. Baluska, M. Gagliano and G. Witzany, 191–213. Cham: Springer.

Simard, S. [W.] 2021. *Finding the Mother Tree: Uncovering the Wisdom and Intelligence of the Forest*. London: Allen Lane.

Singla, A. 2020. 'Dancing Bees Speak in a Code—A Review'. *Emerging Life Science Research* 6(2): 44–53.

Skandrani, Z., M. Desquilbet and A.-C. Prévot. 2018. 'A Renewed Framework for Urban Biodiversity Governance: Urban Pigeons as a Case-Study'. *Natures Sciences Sociétés* 26(3): 280–90

Slade, J. and E. Alleyne. 2023. 'The Psychological Impact of Slaughterhouse Employment: A Systematic Literature Review'. *Trauma, Violence, and Abuse* 24(2): 429–40.

Smeele, S. Q., S. A. Tyndel, L. M. Aplin and M. B McElreath. 2024. 'Multilevel Bayesian Analysis of Monk Parakeet Contact Calls Shows Dialects Between European Cities'. *Behavioral Ecology* 35(1): arad093.

Soga, M. and K. J. Gaston. 2024. 'Global Synthesis Indicates Widespread Occurrence of Shifting Baseline Syndrome'. *BioScience* 74(10) 686–94.

Solomon, D., J. Lehmann, J. A. Fraser et al. 2016. 'Indigenous African Soil Enrichment as a Climate-Smart Sustainable Agriculture Alternative'. *Frontiers in Ecology and the Environment* 14(2): 71–76.

Soma, M. and C. Mori. 2015. 'The Songbird as a Percussionist: Syntactic Rules for Non-Vocal Sound and Song Production in Java Sparrows'. *PLOS One* 10(5): e0124876.

Song, H., 2010. *Pigeon Trouble: Bestiary Biopolitics in a Deindustrialized America*. Philadelphia: University of Pennsylvania Press.

Souris, M., L. Tshilolo, D. Parzy et al. 2022. 'Pre-Pandemic Cross-Reactive Immunity Against SARS-CoV-2 Among Central and West African Populations'. *Viruses* 2022(14): 2259.

Stengers, I. 2010. *Cosmopolitics 1*, translated by R. Bononno. Minneapolis: University of Minnesota Press.

Stirling, A. 2015. 'Emancipating Transformations: From Controlling "The Transition" to Culturing Plural Radical Progress'. In *The Politics of Green Transformations*, edited by I. Scoones, M. Leach and P. Newell, 54–67. London: Routledge.

Strang, V. 2023. *Water Beings: From Nature Worship to the Environmental Crisis*. London: Reaktion Books.
Strathern, M. 1988. *The Gender of the Gift*. Berkeley, CA: University of California Press.
Suzuki, T. N., D. Wheatcroft and M. Griesser. 2016. 'Experimental Evidence for Compositional Syntax in Bird Calls'. *Nature Communications* 2016(7): 10986.
Suzuki, T. N., D. Wheatcroft and M. Griesser. 2017. 'Wild Birds Use an Ordering Rule to Decode Novel Call Sequences'. *Current Biology* 27(15): 2331–36.
Swancutt, K. 2021. 'The Chicken and the Egg: Cracking the Ontology of Divination in Southwest China'. *Social Analysis* 65(2): 40–19.
Swanson, H. A., M. E Lien. and G. B. Ween, eds. 2017. *Domestication Gone Wild: Politics and Practices of Multispecies Relations*. Durham, NC: Duke University Press.
Taiz, L., D. Alkon, A. Draughn et al. 2019. 'Plants Neither Possess Nor Require Consciousness'. *Trends in Plant Science* 24(8): 67–87.
Thiele, K. 2016. 'Quantum Physics and/as Philosophy: Immanence, Diffraction, and the Ethics of Mattering'. *Rhizomes: Cultural Studies in Emerging Knowledge* 30. https://doi.org/10.20415/rhiz/030.e04
Tilly, C. and S. G. Tarrow. 2015. *Contentious Politics*. Oxford: Oxford University Press.
Todd, Z. 2016. 'An Indigenous Feminist's Take on the Ontological Turn: "Ontology" is Just Another Word for Colonialism'. *Journal of Historical Sociology* 29(1): 4–22.
Toutain, A.-G. 2022. 'Sign, Function and Life: Thinking Epistemologically About Biosemiotics'. *Sign Systems Studies* 50(1): 90–132.
Trewavas, A. 2014. *Plant Behaviour and Intelligence*. Oxford: Oxford University Press.
Tronto, J. 2020. *Moral Boundaries: A Political Argument for an Ethics of Care*. New York: Routledge.
Tsegaye, M. 2017. 'Using Ethnography to Understand Gender Dynamics in Poultry Farming'. 14 April. https://www.ilri.org/news/using-ethnography-understand-gender-dynamics-poultry-farming (accessed 26 January 2025).
Tsing, A. 2015. *The Mushroom at the End of the World: On the Possibility of Life in Capitalist Ruins*. Princeton, NJ: Princeton University Press.
Uexküll, J. von. 2010 [1934]. *A Foray into the Worlds of Animals and Humans, with a Theory of Meaning*, translated by J. D. O'Neil. Minneapolis: University of Minnesota Press.
UN DESA [United Nations Department of Economic and Social Affairs]. 2018. '8% of the World Population Projected to Live in Urban Areas by 2050, Says UN'. 16 May. https://www.un.org/development/desa/en/news/population/2018-revision-of-world-urbanization-prospects.html (accessed 27 January 2025).
USDA [United States Department of Agriculture], Natural Resources Conservation Service. n.d. 'Soil Texture Calculator'. https://www.nrcs.usda.gov/resources/education-and-teaching-materials/soil-texture-calculator (accessed 26 January 2025).
Van den Berg, A. E., J. Maas, R. A. Verheij and P. P. Groenewegen. 2010. 'Green Space as a Buffer Between Stressful Life Events and Health'. *Social Science & Medicine* 70(8): 1203–10.
Van Dooren, T. 2014. *Flight Ways: Life and Loss at the Edge of Extinction*. New York: Columbia University Press.
Vine, M. 2019. 'Resonant Herbs: Botanical Socialities, Personal Boundary Work, and the Politics of Self-Care in Southern California'. *Journal of the Royal Anthropological Institute* 25: 285–302.

Viveiros de Castro, E. 2014 [2009]. *Cannibal Metaphysics*, edited and translated by P. Skafish. Minneapolis: University of Minnesota Press.

Walker, J. and M. Nadin. 2011. 'Fruit Bat Cull Urged to Halt Spread of Lethal Hendra Virus'. *The Australian*, 9 July.

Walley, C. 2004. *Rough Waters: Nature and Development in an East African Marine Park*. Princeton, NJ: Princeton University Press.

Warkentin, B. P. 2006. *Footprints in the Soil: People and Ideas in Soil History*. New York: Elsevier.

Warren, K. J. 2000. *Ecofeminist Philosophy: A Western Perspective on What It Is and Why It Matters*. Lanham, MD: Rowman & Littlefield.

Watanabe, S., J. Sakamoto and M. Wakita. 1995. 'Pigeons' Discrimination of Paintings by Monet and Picasso'. *Journal of the Experimental Analysis of Behavior* 63(2): 165–74.

Watene, K. 2024. 'Indigenous Philosophy and Intergenerational Justice' In *Intercultural Philosophy and Environmental Justice Between Generations: Indigenous, African, Asian, and Western Perspectives*, edited by H. Abe, M. Fritsch and M. Wenning, 17–32. Cambridge: Cambridge University Press.

Welchman, A. E. 2016. 'The Human Brain in Depth: How We See in 3D'. *Annual Review of Visual Sciences* 2: 345–76.

Wendelboe-Nelson, C., S. Kelly, M. Kennedy and J. W. Cherrie. 2019. 'A Scoping Review Mapping Research on Green Space and Associated Mental Health Benefits'. *International Journal of Environmental Research and Public Health* 16: 1–49.

Weston, K. 2017. *Animate Planet: Making Visceral Sense of Living in a High-Tech Ecologically Damaged World*. Durham, NC: Duke University Press.

Wheeler, W. 2016. *Expecting the Earth: Life, Culture, Biosemiotics*. London: Lawrence and Wishart.

White, L. 2000. *Speaking with Vampires: Rumor and History in Colonial Africa*. Berkeley, CA: University of California Press.

White, M. P., I. Alcock, J. Grellier et al. 2019. 'Spending at Least 120 Minutes a Week in Nature is Associated with Good Health and Wellbeing'. *Scientific Reports* 9(1): 7730.

Whitehead, H. and L. Rendell. 2015. *The Cultural Lives of Whales and Dolphins*. Chicago: University of Chicago Press.

Whiten, A. 2017. 'A Second Inheritance System: The Extension of Biology Through Culture'. *Interface Focus* 7: 160142.

Whiten, A. 2019. 'Cultural Evolution in Animals'. *Annual Review of Ecology, Evolution, and Systematics* 5: 27–48.

Whiten, A. 2021. 'The Burgeoning Reach of Animal Culture'. *Science* 372: 6537.

Whiten, A., C. A. Caldwell and A. Mesoudi. 2016. 'Cultural Diffusion in Humans and Other Animals'. *Current Opinion in Psychology* 8: 15–21.

Whyte, D. 2019. 'Belonging in the Ocean: Surfing, Ocean Power, and Saltwater Citizenship in Ireland'. *Anthropological Notebooks* 25(2): 13–33.

Winn, P. 2002. 'Everyone Searches, Everyone Finds: Moral Discourse and Resource Use in an Indonesian Muslim Community'. *Oceania* 72(4): 275–92.

Wittig, R. M., C. Crockford, K. E. Langergraber and K. Zuberbühler. 2014. 'Triadic Social Interactions Operate Across Time: A Field Experiment with Wild Chimpanzees'. *Proceedings of the Royal Society B: Biological Sciences* 281: 20133155.

Witzany, G. 2008. 'The Biosemiotics of Plant Communication'. *The American Journal of Semiotics* 24(1–3): 39–56.

Wohlleben, P. 2016. *The Hidden Life of Trees: What They Feel, How They Communicate: Discoveries from a Secret World*. London: Harper Collins.

World Health Organization, 2016. *Urban Green Spaces and Health*. Geneva: World Health Organization Regional Office for Europe.

Worm, M., T. Landgraf and G. von der Emde. 2021. 'Electric Signal Synchronization as a Behavioural Strategy to Generate Social Attention in Small Groups of Mormyrid Weakly Electric Fish and a Mobile Fish Robot'. *Biological Cybernetics* 115(6): 599–613.

Wright, T. F. 1996. 'Regional Dialects in the Contact Call of a Parrot'. *Proceedings of the Royal Society B* 263: 867–72.

Wynne, C.D.L. and M.A.R. Udell. 2020. *Animal Cognition: Evolution, Behavior and Cognition*. London: Red Globe Press.

Yap, M. and E. Yu, 2019. 'Mabu Liyan: The Yawuru Way'. In *Routledge Handbook of Indigenous Wellbeing*, edited by C. Fleming and M. Manning, 261–80. London: Routledge.

Yong, E. 2023. *An Immense World: How Animal Senses Reveal the Hidden Realms Around Us*. New York: Vintage.

Youatt, R. 2020. *Interspecies Politics: Nature, Borders, States*. Ann Arbor, MI: University of Michigan Press.

Zerner, C. 2003. 'Sounding the Makassar Strait: The Poetics and Politics of an Indonesian Marine Environment'. In *Culture and the Question of Rights: Forests, Coasts, and Seas in Southeast Asia*, edited by C. Zerner, 56–108. Durham, NC: Duke University Press.

Zhou, J., J. Lai, G. Menda et al. 2022. 'Outsourced Hearing in an Orb-Weaving Spider that Uses Its Web as an Auditory Sensor'. *PNAS* 119(14): e2122789119.

Zinsstag, J., A. Kaiser-Grolimund, K. Heitz-Tokpa et al. 2023. 'Advancing One Human–Animal–Environment Health for Global Health Security: What Does the Evidence Say?' *The Lancet*, 401 (10376).

Zuberbühler, K. 2012. 'Primate Communication'. *Nature Education Knowledge* 3(10): 83.

Zuberbühler, K., D. Cheney and R. Seyfarth. 1999. 'Conceptual Semantics in a Nonhuman Primate'. *Journal of Comparative Psychology* 113: 33–42.

INDEX

actor-networks, 200
affect, 74–77; forest immersions and, 126–127
agency, 9, 41, 123; actor-networks in, 200; of chickens, 59–60; of forests, 123; of horses, 64, 75, 77; of plants, 87; political naturekind and, 198–203; of trees, 80
Akbar the Great, 45–46
American Beaver and His Works, 26
Anderson, Jon, 152–153
animal culture, 24; companionship in, 50–51
anthropology, 2
arbitrariness, 20–21
Archambault, Julie Soleil, 81
Arndt, Ernst Moritz, 146
Artificial Intelligence, 22, 155
assemblages, 7, 28, 181–182; companionship and, 50–52; forest, 117–118; in the seas, 156, 161; soil, 132–135, 143; urban, 173–174
Augustine of Hippo, 30–31

Ball, Philip, 93
Baptista, João Afonso, 99
Barad, Karen, 8
Barthes, Roland, 33
Barua, Maan, 171, 173–174
Bateson, Gregory, 9, 44
bats, 102–103; sensory sophistication and communication amongst, 103–105; structural biosemiotics of coexistence between humans and, 105–109; synanthropism of, 108; viral entanglements between humans and, 109–113
bees, 90–92; communication between humans and, 94–101; symbolic communication by, 92–94; waggle dances by, 90, 92–93, 96. *See also* insect communication
Bennett, Jane, 8, 165, 200
Biehl, Janet, 146
biodiversity, 1, 89; collapse of, 12; separation between humanised place and spaces for nature, 191–192
biological sciences: need for theory of meaning in, 25–29; revolution in, 18–25, 177
biosemiotics, 3, 4, 9, 17, 26, 29–33, 179–180; bee language, 94; mainstream, 31–33; structural (*see* structural biosemiotics); symbolic signs *versus* indexical/iconic signs in, 32
birds: calls and songs of, 21–23; in cities, 168–172. *See also* chickens
Birke, Lynda, 67
Bourdieu, Pierre, 59
Buscher, Bram, 193

chemical communication, 20
chickens, 189; multimodal communication by and with, 54–57; ontologies and communication across the pluriverse, 61–63; structural biosemiosis in everyday human-chicken life, 57–61; worldwide distribution of, 54
Chomsky, Noam, 46

cities, 163; animals and plants in, 168; birds in, 168–172; communication and health in, 172–176; communicative, 163–166; green urban spaces in, 167–168; urban lives beyond the human and their structural biosemiotics in, 167–172
citizenship theory, 200
Clever Hans, 29, 30, 38, 70
co-becoming, 74–77
codes across species, 38–40
cognition, 35
Colston, Edward, 166
communication, 1–2, 16–17; amidst underwater life, 155–159; anthropology and, 2; arbitrariness in, 20–21; bats, 103–105; beyond the human, 3–5, 178–191; biological revolution and, 18–29; biological revolution transforming, 2; bird songs and calls, 21–23; chemical, 20; chicken-human, 54–57; in cities, 163–166; context in non-human, 23–24; effectiveness of cross-species, 39–40; emplaced, 40–43, 68–74, 87–89; equestrian, 64–68; experiment on innate language and, 17–18; in forests, 115–123; and health in cities, 172–176; houseplant conversation, 80–84; human exceptionalism and, 2, 3, 11, 12, 14, 49, 185, 203; human syntax in, 23; human-bee, 94–101; humans losing the capacity for, 1; implications for multispecies justice, earth law and rights of nature, 195–196; insect, 16, 18–19; as multisensory, 180–181; by plants, 79–80; pluriverse of (*see* pluriverse, the); political naturekind and, 191–197; power in and for, 11–15; and power in and for naturekind, 11–15; prelinguistic signs in, 17, 31–32, 35–36, 37, 83–84, 179; repetitive sounds in, 22; researching beyond-human, 9–11; soils, 132–135; structural analysis of, 15; symbolic, 19, 21, 92; syntax in (*see* syntax); theory of meaning and, 2–3; through electrical signalling, 20; through touch, 19–20; time to develop inter-being, 10–11; in varied ecosystems, 1–2

companionships, 50–52, 187–189; bees and, 91, 98–99; with chickens, 54; entangled with occupation, 51; with horses, 64, 65, 68–72, 74; multispecies, 190; with plants, 78, 81–82; tree, 84–87; violence in, 189–191
Comyn, Marian, 198
connection, ecological, 5–9
context in non-human communication, 23–24
Covid-19 pandemic, 81, 109, 112, 137
Crist, Eileen, 92, 95, 195
critical race and class theory, 196
culture, 3, 4; animal, 24; as second inheritance system, 2, 17, 24, 27–28; theory of meaning for understanding, 25–29
cumulative conditioning, 28

Darwinian evolution, 3, 48, 196
da Vinci, Leonardo, 35–36
Davis, Dona Lee, 69–70
Degnan, Catherine, 84
Deleuze, Gilles, 7, 35
De Montaigne, Michel, 7
Despret, Vinciane, 9, 70
Dobson, J., 175
Douglas, Mary, 42–43
Dracula, 109
Dryzek, John, 201
Duchamp, Marcel, 43

earth law, 195–196
Ebola virus, 109–110
echo-perception, 105–106
ecological connection, 5–9
ecosystems, 28; diversity of, 1–2; marine (*see* seas, the)
electrical signalling, 20
emplaced communication and emplacement, 40–43, 183–184; forests, 123–126; human-horse, 68–74; plants and trees, 87–89; the seas and, 154–155; soil conversations, 143
Enlightenment era, 2, 7, 25, 48
environmental politics, 204

feminist traditions, 195, 200
Fijn, Natasha, 75
Fletcher, Rob, 193
forests, 84–87, 89, 114–115; communication and alignment in flourishing between humans and, 128–130; as communicative communities, 115–123; emplaced, 123–126; immersions and affect, 126–127
Foster, Craig, 157–158
Foucault, Michel, 161, 204
Fraser, Nancy, 198
Frederick II, emperor, 18

Gaia hypothesis, 48
Game, Ann, 75
Ghosh, Amitav, 147, 180, 203
God-given language of humankind, 17–18, 45–46
Gramsci, Antonio, 198
Green Connect urban farming project, 144–145
Greenough, Paul, 39
green public spaces, 167–168
grooming, 19–20
Guattari, Felix, 7

habitus, 59
Haraway, Donna, 7, 50, 203
Hayward, Eva, 156–157
health in cities, 172–176
Helmreich, Stefan, 161
Hinchliffe, Steve, 167, 194, 200
Holmgren, David, 144
horses: Clever Hans, 29, 30, 38, 70; co-becoming and affect of humans and, 74–77; communication by and with, 64–68; personalised and emplaced communication, 68–74; as recurrent feature of social life and history, 64
houseplants. *See* plants
How Forests Think, 3, 32, 115–116
human exceptionalism, 2, 3, 11, 12, 14, 49, 185, 203

immersion in the seas, 150–152
Ingold, Tim, 7–8, 71
innate language, 17–18
insect communication, 16, 18–19; companionship and, 50; termite mounds, 138–139. *See also* bees

James IV, King of Scotland, 17
Jepson, Paul, 194
Jones McVey, Rosalie, 65, 68–69, 74

Kavesh, Muhammed, 170
Kimmerer, Robin, 119–120, 122, 127
kinship with plants, 81
Kockelman, Paul, 62
Kohn, Eduardo, 3–5, 9, 32, 115–116
Kosut, Mary, 97–98
Kull, Kalevi, 26

Lasco, Gideon, 81
Latour, Bruno, 7, 200
Lenin, Vladimir, 198
Lestel, Dominique, 9, 28, 38, 188
Lévi-Strauss, Claude, 4, 34
linguistic communities, 37–40
Lisdorf, Anders, 164
Lovelock, James, 48

Manual of Horsemanship, The, 66–67
Marburg virus, 109–110
Marder, Michael, 80
Marris, Emma, 193
Marx, Karl, 197–199
Maurstad, Anita, 69–70
meaning: communication beyond the human and the power of, 178–191; in emplaced communication, 40–43; packaged, 43–46, 182–183; prelinguistic, 31–32, 35–36, 37, 83–84; social performance and, 34; in study of human social worlds, 34–35; syntax in (*see* syntax); word order and, 33–34. *See also* theory of meaning
Meijer, Eva, 9, 201–202

metasigns: human-soil, 135–136; in soil conversations, 136–143. *See also* signs
metasyntax, 44–45
Miller, Theresa, 81
Mollison, Bill, 144–145
Monbiot, George, 133–134
Montaigne, 48
Moore, Lisa Jean, 97–98
Morgan, Lewis Henry, 7, 26
multispecies justice, 195–196
mutual flourishing, human-soil communication, 144–146
My Octopus Teacher, 157

narratives, human-soil, 135–136
natural signs, 30–31
Nazism, 146–147

O'Brien, Anne Therese, 146
Occam's razor, 26, 64
Other Effective Area-based Conservation Measures (OECMs), 191–192

packaged meanings, 43–46, 182–183
Pálsson, Gisli, 161
Peirce, Charles, 4, 32, 37
perception, 35, 41–42
PerfectBee, 95–96
permaculture, 145–146
Phillips, Catherine, 80
philosophies of mind, 42
Pickering, Jon, 201
pigeons, urban, 168–172
Piva, Heidi Campana, 96–97
place, interpretation of, 40–41
plants, 78; in cities, 168; communicative, 79–80; emplaced communication, 87–89; houseplant conversations, 80–84; kinship with, 81; social learning in, 25. *See also* trees
pluriverse, the: companionships and assemblages in, 50–52; emplaced plants and trees in, 88–89; framing chicken ontologies and communication across, 61–63; human-bat communication in, 107–108; human health in, 113; respect for Indigenous worlds and, 194; the seas and, 151; structural biosemiotics across, 47–49

Polanyi, Karl, 198
political naturekind, 191–197; process of doing, 197–204
prelinguistic signs, 17, 31–32, 35–36, 37, 83–84, 179; in plants, 83–84
Puig de la Bellacasa, Maria, 132–133, 137, 143, 145, 195
Purity and Danger, 42
Pythagoras, 7, 49, 190

Ramasawmy, Melanie, 62–63
relational materialities, 132
religion and spirituality, 6, 14; companionship and, 50; reincarnation in, 190; surfing and, 153
Rembrandt, 35
repetitive sounds, 22
rights of nature, 195–196
Rosevear, D. R., 126

Safina, Carl, 13, 24
Saussure, Ferdinand de, 31
Schulz, Eily, 80
Sea Change Project, 157–158
seas, the, 149; communication amidst underwater life in, 155–159; elemental communications and immersions in, 150–152; emplaced narratives of, 154–155; negotiation and respect in entanglement between humans and, 159–162; surfing conversations and, 152–154
Sebeok, Thomas, 4, 31–32, 90
Seeley, Thomas, 91
semiosis, 3, 10; bats, 103–105; prelinguistic, 31–32
semiotics, 29–33; in antiquity, 30–31
separations, human-nature, 191–193
Shareable, 164
Sharpless, Ike, 203

Sheldrake, Merlin, 133
shifting baseline syndrome, 13
signals, 19, 27; chemical, 20; electrical, 20; order of, 21; packaged meaning, 44–45; vocal, 21
signifiers and signified, link between, 36
signs: emplaced communication, 42–43; human-plant communication, 82–83; natural, 30–31; packaged meaning, 43–44, 182–183; prelinguistic, 17, 31–32, 35–36, 37, 83–84, 179; structural dimensions to, 36–37. *See also* metasigns
Simard, Suzanne, 79, 116
smart systems, 175–176
social learning, 24–25, 27–28; bees and, 93
social performance and syntax, 34
soils, 131–132; assemblages of, as communicative communities, 132–135; conversations and significance of metasigns and narratives between humans and, 135–136; emplaced conversations, 143; land, and entanglements of identity, 146–148; metasigns in conversations of, 136–143; mutual flourishing in conversations between humans and, 144–146
Spinoza, Baruch, 7, 48
spirituality. *See* religion and spirituality
Staudenmeier, Peter, 146
Stengers, Isabelle, 8, 200
Stoker, Bram, 108–109
Strang, Veronica, 151, 160, 203
StrongTowns, 164
structural biosemiotics, 33–49, 177; across the pluriverse, 47–49; in cities, 174–176; dimensions to signs in, 36–37; emplaced communication, 40–43; in everyday human-chicken life, 57–61; human-bat coexistence, 105–113; iconic and indexical images in, 35–36; linguistic communities, 37–40; packaged meanings, 43–46; plants and, 82; on prelinguistic language, 35–36, 37; signifier and signified linked in, 36; on social performance, 34; in study of human social worlds, 34–35;

theory of language, perception, and cognition in, 35, 41; urban lives beyond the human and their, 167–172; on word order, 33–34
surfing conversations, 152–154
Swancutt, Katherine, 61
symbolic communication, 19, 21; by bees, 92
syntax, 19, 21, 27, 182, 184; emplaced communication, 42–43; in human language, 23; mainstream biosemiotics on, 31; social performance and, 34

tactile communication, 19–20
Taylor, Hollis, 188
termite mounds, 138–139
theory of meaning, 2–3; in biology, 25–29; linguistic communities and, 37–40. *See also* meaning
theory of mind, 29
Thompson, Kirrilly, 67
touch, communication through, 19–20
trans-specific communication, 28
Treadwell, Timothy, 39–40
trees: communicative companionships and historical entanglements, 84–87; emplaced communication, 87–89. *See also* plants
Tronto, Joan, 195
tropism, 2, 25
Tsing, Anna, 191

umwelt, 38–39, 60–61, 102, 104, 158–159, 168
United Nations Declaration of Human Rights, 196
United Nations Food and Agriculture Organization (FAO), 114, 144
urban pigeons, 168–172

Vine, Michael, 81
violence in companionship, 189–191
viral entanglements, human-bat, 109–113
von Frisch, Karl, 19

waggle dances, 90, 92–93, 96
Weston, Kath, 191
Whatmore, Sarah, 167, 194, 200
Whyte, David, 152
Wilding Network, 137
Winn, Philip, 147

Wohlleben, Peter, 79, 116
word order, 33–34

Xavier, Hieronymus, 46

Zerner, Charles, 157

GPSR Authorized Representative: Easy Access System Europe - Mustamäe tee
50, 10621 Tallinn, Estonia, gpsr.requests@easproject.com

www.ingramcontent.com/pod-product-compliance
Ingram Content Group UK Ltd.
Pitfield, Milton Keynes, MK11 3LW, UK
UKHW042216300925
463475UK00001B/5